Surveys and Tutorials in the Applied Mathematical Sciences

Volume 16

Featuring short books of approximately 80-200pp, Surveys and Tutorials in the Applied Mathematical Sciences (STAMS) focuses on emerging topics, with an emphasis on emerging mathematical and computational techniques that are proving relevant in the physical, biological sciences and social sciences. STAMS also includes expository texts describing innovative applications or recent developments in more classical mathematical and computational methods.

This series is aimed at graduate students and researchers across the mathematical sciences. Contributions are intended to be accessible to a broad audience, featuring clear exposition, a lively tutorial style, and pointers to the literature for further study. In some cases a volume can serve as a preliminary version of a fuller and more comprehensive book.

Vakhtang Putkaradze

A Concise Introduction
to Classical Mechanics

 Springer

Vakhtang Putkaradze
Department of Mathematical and Statistical
Sciences
University of Alberta
Edmonton, AB, Canada

ISSN 2199-4765   ISSN 2199-4773   (electronic)
Surveys and Tutorials in the Applied Mathematical Sciences
ISBN 978-3-031-84976-3   ISBN 978-3-031-84977-0   (eBook)
https://doi.org/10.1007/978-3-031-84977-0

Mathematics Subject Classification: 70G75, 70-01, 70-03, 70G45, 70H05

This Springer imprint is published by the registered company Springer Nature Switzerland AG
The registered company address is: Gewerbestrasse 11, 6330 Cham, Switzerland

If disposing of this product, please recycle the paper.

Preface

Why This Book?

There are many excellent books on classical mechanics available today. So, the reader may ask, why write another one? This set of lecture notes was written for a course in Classical mechanics for Math Physics majors that I was teaching at the University of Alberta. When I set forth writing these lecture notes, my goal was to formulate the modern ideas of classical mechanics in a language that is understandable to undergraduates and yet gives a hint of tools developed for and used in contemporary classical mechanics. The fundamental tools based on geometry are hardly discussed here; however, I hope this set of notes will be helpful to the readers who become interested enough in the subject to undertake further reading.

What did I want to achieve with these lecture notes? First and foremost, I want to show that classical mechanics is an active, living, and breathing subject. The mechanics itself is indispensable to many areas of science and engineering, from mathematical biology to nanotechnology, engineering, and many other fields, even machine learning. The modern concepts of classical mechanics, such as dynamics on manifolds, symmetry invariance, variational principles, and others, can unify many problems that seemingly have very little to do with each other. As discussed in this book, the methods developed for and inspired by classical mechanics find applications in many fields, such as engineering, economics, machine learning, and many others. To facilitate this discussion, every chapter in the book has a section called *Looking Forward*, where I briefly discuss recent applications of knowledge presented in that particular chapter.

The book is primarily dedicated to classical mechanics taught from variational principles. In the introductory mechanics courses, a reader may have encountered Newton's laws of physics and was taught to solve mechanics problems by balancing forces and torques on a system. While it is true that every problem in mechanics can be solved that way, not every problem should. For many problems having complex interacting parts, balancing such forces and even keeping track of all of them may be difficult. Variational principles provide an equivalent and a much more systematic way of writing equations of motion but give insights into the additional structure of solutions, even before any solution is sought.

The book is aimed at a typical one-semester, three-credit course for undergraduates who study mathematics, physics, engineering, or other areas where the ideas of analytical mechanics may be useful. A typical student in the course had preparation in Calculus; they would also have taken an introductory mechanics course. They would have, as a rule, no preparation in geometric approaches to mechanics or physics. Also, I do not assume knowledge of linear algebra—the exposition of that topic is self-contained in the book.

The book contains all the information from a standard list of topics in mechanics that is feasible to teach during a one-semester course. The book is meant to be *concise* and easy to read, and thus follows the following principles:

- The text includes only the core part of the course essential for analytical mechanics that is realistic to give during one semester.
- Mathematical ideas used in modern mechanics behind the methods are emphasized.
- Just enough mathematical rigor is introduced to develop concepts of modern mechanics without over-burdening the reader with technical details.
- The citations to other sources are reduced to the minimum and are only used when there is a chance that an interested reader would develop enough interest to continue reading beyond the material provided in the text. The exceptions to this rule are the *Looking Forward* sections, where I try to show how the concepts developed in each particular section are useful in modern science and applications.
- The book presents a reasonable (but not excessive) number of solved examples. Solutions to the selected homework problems are presented in Appendix A.
- The language is intentionally kept light for the ease of use by a wide community of students and researchers.
- There is a substantial number of solved examples in the text.
- Every chapter has several practice problems to test students' knowledge of the material. These problems, or problems of similar difficulty, were given as homework in the class. Appendix A also gives detailed solutions to the practice problems presented in the text.
- The book presents a sample of midterm and final exams for students to practice.

In preparation for these notes, I have consulted the fundamental book by V. I. Arnol'd [1], modern exposition of mechanics by D. D. Holm [2, 3], the modern classic book by J. Marsden and T. Ratiu [4], and an excellent book by J. V. Jose and E. J. Saletan [5]. I hope that the present book will serve as a bridge from the excellent textbooks in mechanics based on the traditional approaches such as Landau and Lifshitz [6] and Sommerfeld [7] to these more modern monographs exposing the beauty of mechanics using geometric considerations. I also hope that my book will go beyond being a simple bridge between the two expositions by illustrating how modern geometric methods in mechanics yield important results. By the end of the course, the reader should be convinced that the modern methods of geometry are not just beautiful but also useful for solving practical problems.

Several exercises for this course were inspired by an excellent collection of problems contained in [8], which is, alas, only accessible in Russian at this point. Many problems from that collection were also used by students in preparation for the exams. I hope that, eventually, a translation of this extraordinary, 400-page-long collection of problems will be undertaken for the benefit of both the students and the instructors in classical mechanics worldwide.

Acknowledgments

I have drawn inspiration for this book from many years of collaboration with my colleagues and friends Profs. A. M. Bloch (University of Michigan), F. Gay-Balmaz (NTU Singapore), M. Leok (UCSD), P. Olver (University of Minnesota) T. S. Ratiu (EPFL/SJTU), C. Tronci (University of Surrey), and many others. I am particularly grateful to Prof. D. D. Holm (Imperial College), who inspired me to join the field of geometric mechanics several decades ago through his passion for the subject. Throughout my career, he was not just my long-term colleague and collaborator but also a mentor and a good friend. It was my privilege to have met and held discussions with Prof. J. E. Marsden, who was, and still is, an inspiring and towering figure in the field. I am also indebted to Prof. D. V. Zenkov (NCSU), who provided many insightful and critical comments about the pedagogical exposition of various aspects of mechanics in a clear and modern way.

I am grateful to the students of MaPh343, the classical mechanics course for Math Physics students at the University of Alberta, who were diligent in finding errors and mistakes and helped improve the exposition in these notes. I am also grateful to Sofiia Huraka, who read the manuscript very carefully and provided numerous corrections and suggestions. In addition, I would like to acknowledge the support and discussions with my colleagues at the University of Alberta. I am also indebted to the inspiring and creative atmosphere in the ATCO Ltd Transformation Team, where I had the privilege to work for several years.

Finally, I would like to thank George Putkaradze for providing many of the light-hearted yet artistic and scientifically accurate illustrations for this book.

Edmonton, AB, Canada   Vakhtang Putkaradze

Contents

1 The Newton's Laws of Motion and Conservation Laws 1
   1.1   The Newton's Laws of Motion 1
   1.2   Linear and Angular Momentum: Single-Particle Case 3
   1.3   Many-Particle Systems: Angular and Linear Momenta 4
   1.4   Problems Involving Changing Mass 6
   1.5   Looking Forward 7
   1.6   Practice Problems 8

2 From Newton's Laws to Euler-Lagrange Equations 11
   2.1   Euler-Lagrange Equations 11
   2.2   Energy Definition and Conservation 14
   2.3   Motion in Central Potential and Kepler's Law 16
   2.4   Looking Forward 21
   2.5   Practice Problems 21

3 Configuration Manifolds, Variational Principle, and Euler-Lagrange Equations 23
   3.1   Preliminaries: Notation and Definitions 23
   3.2   Manifolds and Tangent Bundles 26
   3.3   Configuration Manifolds 28
   3.4   Tangent Bundle of a Smooth Manifold 29
   3.5   Action and Variations 29
   3.6   Hamilton's Critical Action Principle 30
   3.7   Worked Examples 32
   3.8   Holonomic Constraints 38
   3.9   Lagrange-d'Alembert's Principle for External Forces and Rayleigh's Dissipation Function 40
   3.10   Looking Forward 42
   3.11   Practice Problems 42

4   Noether's Theorem and Conservation Laws 45
    4.1   Noether's Theorem 45
    4.2   Examples.. 47
    4.3   Invariance of Dissipative Systems 49
    4.4   Looking Forward 49
    4.5   Practice Problems... 50

5   Linear Stability of Small Oscillations About an Equilibrium 53
    5.1   A Review in Linear Algebra 53
    5.2   Linear Stability of a Mechanical System 56
    5.3   Examples.. 59
    5.4   Looking Forward 63
    5.5   Practice Problems... 63

6   Hamiltonian Systems.. 65
    6.1   Derivations of Hamilton's Equations 65
    6.2   Examples of Dynamics in Hamiltonian Representation 67
    6.3   Cotangent Bundle... 70
    6.4   Poisson Brackets... 71
    6.5   Examples of Poisson Bracket Computations 74
    6.6   Looking Forward 75
    6.7   Practice Problems... 75

7   Introduction to Differential Forms and Exterior Calculus 79
    7.1   Vector Fields, One-Forms, and Exterior Products................. 79
    7.2   Pullback of Differential Forms 84
    7.3   Applications to Hamiltonian Mechanics........................... 85
    7.4   Integration of Differential Forms: Stokes Formula and
          Integral Invariants of Poincaré and Poincaré-Cartan 86
    7.5   Looking Forward 87
    7.6   Practice Problems... 88

8   Canonical Transformations... 89
    8.1   General Properties 89
    8.2   Generating Functions of Canonical Transformations 94
    8.3   Darboux' and Liouville's Theorem and Conservation
          of Phase Volume 99
    8.4   Looking Forward 101
    8.5   Practice Problems... 102

9   Hamilton-Jacobi Equation.. 103
    9.1   Derivation of the Hamilton-Jacobi Equation 103
    9.2   Separable Systems 104
    9.3   Action-Angle Variables... 110
    9.4   Adiabatic Invariants 112
    9.5   Looking Forward 114
    9.6   Practice Problems... 114

10 Rigid Body Dynamics 117
    10.1 Motion of a Rigid Body and Rotation Matrices 117
    10.2 Description of Rotations of a Rigid Body Using Matrices 118
    10.3 Body and Spatial Angular Velocities 120
    10.4 Euler's Equations of Motion for a Rigid Body 122
    10.5 Euler's Equations for the Rigid Body Motion 124
    10.6 Heavy Top 124
    10.7 Solution for the Lagrange Top 126
    10.8 Looking Forward 128
    10.9 Practice Problems 129

11 Nonholonomic Constraints 131
    11.1 Constraints and Their Validity 131
    11.2 Lagrange-d'Alembert's Principle 132
    11.3 Examples of Systems with Non-holonomic Constraints 133
    11.4 Looking Forward 138
    11.5 Practice Problems 138

12 Euler-Poincaré Variational Theory for a Rigid Body 141
    12.1 Symmetry of Mechanical Systems 141
    12.2 Notations and Definitions 142
    12.3 Variational Derivation of Rigid Body Equations 143
    12.4 Looking Forward 145

13 Sample Midterm and Final Exams 147
    13.1 Midterm 1 147
    13.2 Midterm 2 149
    13.3 Final Exam 151

A Solutions to Selected Practice Problems 159
    A.1 Chapter 1 159
    A.2 Chapter 2 160
    A.3 Chapter 3 162
    A.4 Chapter 4 174
    A.5 Chapter 5 176
    A.6 Chapter 6 180
    A.7 Chapter 7 184
    A.8 Chapter 8 187
    A.9 Chapter 9 189
    A.10 Chapter 10 195
    A.11 Chapter 11 197

Bibliography 201

Index 207

Chapter 1
The Newton's Laws of Motion and Conservation Laws

1.1 The Newton's Laws of Motion

You have probably seen Newton's laws of motion. They have formed the foundation of classical mechanics for hundreds of years, and we cannot start our discussion without them. Our discussion here is close to the axiomatics of mechanics in [2, 5]. To formulate the laws, we need to define the concept of an isolated particle.

Definition 1.1 An isolated particle is a material object that has no interaction with any other object.

This definition is somewhat non-rigorous as we do not define what "interaction" actually means. We assume that the concept of interaction can be defined through, for example, the existence of external forces acting on a particle. This concept is actually a bit tricky: to verify the existence of external forces, one needs to measure such forces. We shall not delve too much into this subject and simply assume that external forces on a particle can be defined.

The First Newton's Law There exist coordinate frames, called *inertial frames*, such that any isolated particle in that frame will move in a straight line with constant velocity.

The Second Newton's Law If, in such an inertial frame, there is a force **F** acting on the body, then there exists a constant m characteristic of this particular body, called the mass, such that the acceleration of the particle is proportional to **F**, so

$$m\dot{\mathbf{v}} = m\mathbf{a} = \mathbf{F}, \quad \Leftrightarrow \quad ma^i = m\ddot{x}^i = F^i, \quad i = 1, 2, 3. \tag{1.1}$$

The dot above a quantity, here and everywhere else in this book, denotes the time derivative of that quantity, two dots denote the second time derivative, and so forth. Before we venture into the discussion of the third law, let us talk about interactive

© The Author(s), under exclusive license to Springer Nature Switzerland AG 2025
V. Putkaradze, *A Concise Introduction to Classical Mechanics*,
Surveys and Tutorials in the Applied Mathematical Sciences 16,
https://doi.org/10.1007/978-3-031-84977-0_1

particle systems. Suppose we have a system of particles, $i = 1, \ldots, N$, which experiences no external force. Each particle i within this set experiences the force $\mathbf{F}_i$. Because the system is isolated, the only source of the force must be coming from other particles in the system. If we denote the force exerted by the particle j on the particle i as $\mathbf{f}_{ij}$, we necessarily have

$$\mathbf{F}_i = \sum_{j=1}^{N} \mathbf{f}_{ij} \,. \tag{1.2}$$

It is clear that the net force acting on the system must vanish since the system is isolated and the coordinate frame is inertial. Thus, the sum of all forces $\mathbf{F}_i$ in (1.2) for all $i = 1, \ldots, N$ must vanish:

$$\sum_{i=1}^{N} \mathbf{F}_i = \sum_{i,j=1 \,, i \neq j}^{N} \mathbf{f}_{ij} = \mathbf{0} \,. \tag{1.3}$$

Equation (1.3) gives a condition for the net sum of forces to vanish. Newton's third law actually tells that the cancellation of forces occurs through the antisymmetry $\mathbf{f}_{ij} = -\mathbf{f}_{ji}$.

The Third Newton's Law In an inertial frame, for an isolated system of N particles, $\mathbf{f}_{ij} = -\mathbf{f}_{ji}$ in Eq. (1.3), also known as *For every action, there is an equal and opposite reaction.*

In other words, the force exerted by the particle j on the particle i is equal in magnitude and has the opposite sign to the force exerted by the particle i on the particle j. Clearly, for this choice of forces, Eq. (1.3) is satisfied. However, it is just one way of cancelling the forces. Newton's third law says that this is the only way the cancellation of forces in Eq. (1.3) happens in mechanics.

Discussion You have most likely encountered these three laws before, in the form that they are formulated above—or, perhaps, something close to it. However, when you start thinking about these laws as they are presented, some things become a bit confusing. For example, why can't we redefine time to, say, $t \to t^2$? In mechanics formulated in terms of that new time, there will be no inertial frames as the constant velocity becomes nonconstant. Why would a choice of time like ours exists? Also, what is the force, and how is it defined exactly? Why is mass a constant, and why can't it depend on the choice of the inertial frame (actually, it does in relativistic formulation)? These questions are easy to dismiss as irrelevant. However, some of the greatest minds of the last century were trying to find an axiomatic exposition of these laws. You can read more about these axioms in, for example, excellent books of Jose and Saletan [5] or Holm [2].

1.2 Linear and Angular Momentum: Single-Particle Case

Let us now define the relevant physical quantities for one particle:

1. *Linear momentum* (sometimes just called momentum) $\mathbf{p} = m\mathbf{v}$. *If the mass of the particle is constant*, the second Newton's law is then written as

$$\dot{\mathbf{p}} = \mathbf{F}. \tag{1.4}$$

   Equation (1.4) is known as the *conservation of the linear momentum.*

2. *Angular momentum* $\mathbf{l} = \mathbf{x} \times \mathbf{p}$. Here, $\mathbf{x}$ is a vector connecting a fixed point—the origin—to the particle. Then,

$$\dot{\mathbf{l}} = \dot{\mathbf{x}} \times \mathbf{p} + \mathbf{x} \times \dot{\mathbf{p}} = \underbrace{\mathbf{v} \times m\mathbf{v}}_{=0} + \mathbf{x} \times \mathbf{F} = \mathbf{T}, \tag{1.5}$$

   where $\mathbf{T}$ is the torque acting on the particle. Equation (1.5) is known as the *conservation of the angular momentum.*

3. *Work and kinetic energy* For a trajectory $\mathbf{x}(s)$ starting at t_0, consider the following line integral W_C, along the trajectory $C : \mathbf{x}(s)$, starting at $\mathbf{x}(t_0) = \mathbf{x}_0$ ending at $\mathbf{x}(t)$, for $t_0 \leq s \leq t$:

$$W_C = \int_{t_0}^{t} \mathbf{F} \cdot d\mathbf{x}(s) = \int_{t_0}^{t} m\ddot{\mathbf{x}}(s) \cdot \dot{\mathbf{x}}(s) ds$$
$$= \frac{1}{2} m \int_{t_0}^{t} \frac{d}{ds} |\dot{\mathbf{x}}(s)|^2 ds = \frac{1}{2} m |\mathbf{v}|^2 - \frac{1}{2} m |\mathbf{v}_0|^2. \tag{1.6}$$

   The expression W_C defined by the first integral of the left-hand side of Eq. (1.6) is called work performed by the force $\mathbf{F}$ along the trajectory $\mathbf{x}(s)$. The result (Eq. (1.6)), connecting the work to the kinetic energy $T = \frac{1}{2} m |\mathbf{v}|^2$, is known as the *work-energy theorem* and is sometimes written as $W(t) = T(t) - T(t_0)$. Notice that the result (Eq. (1.6)) is not related to what the functional form of the force $\mathbf{F}$ may be. The work-energy theorem (Eq. (1.6)) is true for all forces, whether they are potential (i.e., expressed as a gradient of some potential, also known as conservative), caused by friction or being caused by any other physical effects.

4. *Energy conservation* Suppose now that the force $\mathbf{F}$ is potential, so $\mathbf{F} = -\frac{\partial U}{\partial \mathbf{x}}$. The work integral in Eq. (1.6) simplifies as

$$W = -\int_{t_0}^{t} \frac{\partial U}{\partial \mathbf{x}} \cdot d\mathbf{x}(s) = U(\mathbf{x}_0) - U(\mathbf{x}) = T(\mathbf{x}) - T(\mathbf{x}_0), \quad \Rightarrow$$
$$E = T(\mathbf{x}_0) + U(\mathbf{x}_0) = T(\mathbf{x}) + U(\mathbf{x}) = \text{const}, \tag{1.7}$$

and the work is independent of the path taken in the $\mathbf{x}$ space from $\mathbf{x}_0$ to $\mathbf{x}(t)$. The result (Eq. (1.7)) is known as energy conservation for the quantity $E = T + U$. As we shall learn in this course, the law of conservation of energy (Eq. (1.7)) is much deeper than meets the eye.

1.3   Many-Particle Systems: Angular and Linear Momenta

Suppose now we have a set of N interacting particles. The exact nature of interactions (springs, rubber bands, sticky gooey threads, intertwined grass, tiny bugs clutching at each other's arms etc.) is not important. The results will be valid *no matter what the interaction is*. However, one should remember that the nature of the interaction is important for the computation of the trajectories of the individual particles. Here, we shall talk about the quantities that can be computed for *any* interaction of the particles.

Linear Momentum   Choose the particles i and $j \neq i$. By Newton's third law, the force $\mathbf{f}_{ij}$ acting on the particle i from j must be equal and opposite to the force acting on the particle j from i, $\mathbf{f}_{ij} = -\mathbf{f}_{ji}$. The particle self-interaction, i.e., $\mathbf{f}_{ii}$, must vanish—if it is not zero, all kinds of bad things (or, perhaps, unrealistically good things) will happen in the world. Suppose there is also some kind of external force $\mathbf{F}_{e,i}$ acting on the particle i. We define the total momentum $\mathbf{P} = \sum_i \mathbf{p}_i$ and compute the evolution of this quantity as

$$\dot{\mathbf{p}}_i = m_i \ddot{\mathbf{x}}_i = \sum_j \mathbf{f}_{ij} + \mathbf{F}_{e,i} \quad \Rightarrow$$

$$\dot{\mathbf{P}} = \sum_i m_i \ddot{\mathbf{x}}_i = \underbrace{\sum_{i \neq j} \mathbf{f}_{ij}}_{=0} + \sum_i \mathbf{F}_{e,i} \quad \Rightarrow \tag{1.8}$$

$$M\ddot{\mathbf{X}} = \mathbf{F}_e , \quad \mathbf{F}_e := \sum_i \mathbf{F}_{e,i} \quad \text{(sum over all external forces)}$$

Here, we defined the total mass and the position of the center of mass as

$$M := \sum_i m_i , \quad \mathbf{X} := \frac{1}{M} \sum_i m_i \mathbf{x}_i . \tag{1.9}$$

In particular, if there are no external forces, or the sum of external forces on the system vanishes, the quantity $\mathbf{P} = M\dot{\mathbf{X}}$ is constant:

$$\sum_i \mathbf{F}_{e,i} = \mathbf{0} \quad \Rightarrow \quad \mathbf{P} = M\dot{\mathbf{X}} = \text{const}, \tag{1.10}$$

which is also known as the *conservation of net linear momentum*. The meaning of Eq. (1.8) is the following: to compute the evolution of the center of mass, one needs to sum all the external forces acting on the ensemble of particles. That is actually really helpful—had it not been the case, instead of computing the motion of a body (say, a rock thrown in the air), we would have to compute the forces and motion of all the atoms interacting within the body. Clearly, such a complexity would contradict common sense and, fortunately, the fundamental principles of mechanics.

Angular Momentum Let us define the quantity $\mathbf{l}_i = \mathbf{x}_i \times \mathbf{p}_i$ to be the angular momentum of the i-th particle. To make further progress, let us introduce the coordinates $\mathbf{y}_i$ computed as the deviation of $\mathbf{x}_i$ from the center of mass $\mathbf{X}$, i.e.: $\mathbf{x}_i = \mathbf{X} + \mathbf{y}_i$. We have, by definition of $\mathbf{X}$,

$$M\mathbf{X} = \sum_i m_i \mathbf{x}_i = \sum_i m_i (\mathbf{X} + \mathbf{y}_i) = \mathbf{X}\sum_i m_i + \sum_i m_i \mathbf{y}_i, \quad \Rightarrow \quad \sum_i m_i \mathbf{y}_i = \mathbf{0}. \tag{1.11}$$

Then, the time derivative of the total angular momentum $\mathbf{L} = \sum_i \mathbf{l}_i$ is

$$\frac{d\mathbf{L}}{dt} = \sum_i m_i \left(\mathbf{x}_i \times \ddot{\mathbf{x}}_i + \underbrace{\dot{\mathbf{x}}_i \times \dot{\mathbf{x}}_i}_{=0} \right) = \sum_i m_i (\mathbf{X} + \mathbf{y}_i) \times (\ddot{\mathbf{X}} + \ddot{\mathbf{y}}_i)$$

$$= \mathbf{X} \times M\ddot{\mathbf{X}} + \sum_i m_i \mathbf{y}_i \times \ddot{\mathbf{y}}_i + \underbrace{\left(\sum_i m_i \mathbf{y}_i \right) \times \ddot{\mathbf{X}} + \mathbf{X} \times \sum_i m_i \ddot{\mathbf{y}}_i}_{= \mathbf{0} \text{ due to } (1.11)} \tag{1.12}$$

$$= \mathbf{X} \times M\ddot{\mathbf{X}} + \sum_i \mathbf{y}_i \times m_i \ddot{\mathbf{y}}_i = \mathbf{X} \times \mathbf{F}_{\text{tot}} + \sum_i \mathbf{y}_i \times \mathbf{F}_i,$$

where we have defined the total force $\mathbf{F}_{\text{tot}} = \sum_i \mathbf{F}_i$. In particular, *in the absence of external forces*, that total force $\mathbf{F}_{\text{tot}}$ vanishes. Moreover, we can choose an inertial frame for which the center of mass is fixed at $\mathbf{X} = 0$. The angular momentum, measured in this frame, is denoted $\mathbf{L}_c = \sum_i \mathbf{y}_i \times m_i \dot{\mathbf{y}}_i$. Equation (1.12) gives the expression for the evolution of $\mathbf{L}_c$ as

$$\frac{d\mathbf{L}_c}{dt} = \sum_{i,j} \mathbf{y}_i \times \mathbf{f}_{ij} = (\text{split up for } i < j \text{ and } i > j)$$

$$= (\text{interchange } i \text{ and } j) = \sum_{i<j} (\mathbf{y}_i - \mathbf{y}_j) \times \mathbf{f}_{ij} = \mathbf{T}, \tag{1.13}$$

where $\mathbf{T}$ is the net torque acting on the system. If, $\mathbf{T} = 0$, which may happen, for example, for $\mathbf{f}_{ij} \parallel (\mathbf{y}_i - \mathbf{y}_j)$, then $\mathbf{L}_c =$const.

1.4   Problems Involving Changing Mass

We shall talk briefly here about systems with changing mass to get a better feel for what Newton's laws mean in terms of physics. Also, the problems of changing mass present a case when it is easy to reach erroneous conclusions by a careless application of variational methods—which is the main topic of this book (the correct application of variational methods will still work). To illustrate this concept, consider the following deceptively simple problem.

Suppose a cart of mass m_0 moves with the speed v_0 on a straight line with no friction. Then, clearly, you will have $v_0 =$const due to Newton's first law. Suppose at time $t > t_0$ the mass of this cart starts to change; more precisely, we are adding mass $m(t)$ to it by, say, pouring sand on the cart, as shown in Fig. 1.1. The functional dependence $m(t)$ is fixed and is independent of velocity, and we would like to find $v(t)$ at a given time. Then, from the conservation of linear momentum, you would like to write

$$\dot{P} = \frac{d}{dt}\left[(m_0 + m(t))\,v(t)\right] = 0\,(?) \quad \rightarrow$$

$$(m_0 + m(t))\,v(t) = m_0 v_0\,(?) \quad \rightarrow \tag{1.14}$$

$$v(t) = \frac{m_0}{m_0 + m(t)} v_0\,(?).$$

Fig. 1.1 Sand being added to a moving cart

Why did we write question marks? If you think about this problem a bit more, you realize that the solution (Eq. (1.14)) is, in general, incorrect. More precisely, the problem, as it is formulated, is impossible to solve since not all relevant information was given. Let us formulate—and solve—this problem more precisely.

Let us say that during the time interval $[t, t + \Delta t]$, the mass added to the cart is Δm. We also assume that the cart is moving with the velocity v in the given inertia frame. The sand added to the cart moves with some velocity with respect to the center of mass $V(t)$. Let us consider two time points: before and after the impact of the sand onto the cart:

1. Before the sand hits the cart during this time interval, the momentum of the sand is $V \Delta m$, and the momentum of the cart is $M(t)v$ where $M = m_0 + m(t)$. The sand and the cart are isolated from each other, and thus, their momentum does not change.
2. After the sand hits the cart, there are interaction forces between the cart and the sand, equilibrating their velocity, so after the impact, the velocity of the sand and the cart is the same. Let us call that velocity $v + \Delta v$. Since all the forces are internal, the total momentum is conserved. We have, therefore,

$$(M + \Delta m)(v + \Delta v) = Mv + V \Delta m, \quad \Rightarrow \quad M \Delta v + \Delta m (v + \Delta v) = V \Delta m, \tag{1.15}$$

which, in the limit of small $\Delta m \simeq \dot{m}(t) \Delta t$ and $\Delta v \simeq \dot{v}(t) \Delta t$, becomes

$$\big(m_0 + m(t)\big)\dot{v}(t) + \dot{m}(t)v = V(t)\dot{m}(t), \quad \Rightarrow \quad \frac{d}{dt}\Big[\big(m_0 + m(t)\big)v\Big] = V(t)\dot{m}(t). \tag{1.16}$$

Now, it is clear that Eq. (1.4) is a particular case of Eq. (1.15) when the sand is added with $V = 0$, in other words, with zero momentum. If the sand has a nonzero velocity when hitting the cart, a more complex equation (Eq. (1.15)) will need to be solved to find $v(t)$ given the rate of mass addition $\dot{m}(t)$ *and* the velocity of the added sand $V(t)$.

1.5 Looking Forward

Newton's method was an extraordinary breakthrough in science, facilitating quantitative change in how we describe physical phenomena. Because of Newton, we have learned how to describe nature using differential equations. Powerful as Newton's method is, it is often difficult to use this method for systems that, for example, consist of several interacting moving parts. The rest of this text will deal with the methods that are alternative but just as powerful as Newton's methods when applied to problems of mechanics. These methods were first developed by Euler and Lagrange and are often referred to as *analytical mechanics*.

However, Newton's method is still superior when considering the motion of systems with changing mass. The most important application of such systems is rocket propulsion. To accurately compute the momenta and torques generated by jets, one still uses some version of Tsiolokovsky's formulas; see Problem 3 in this chapter [9–11]. If you will continue your professional journey into some aspects of rocket and space engineering, such as engine design, knowledge of systems with changing mass will be useful. In many other applications of space research, the theories we derive later in this book will be very useful as well.

1.6 Practice Problems

1. Find a mechanical system which is not isolated, i.e., is acted upon by an external force, for which $\mathbf{F}_{tot} = \sum_i \mathbf{F}_{e,i} = \mathbf{0}$, but the angular momentum taken with respect to the center of mass is conserved.
2. Suppose $m(t)$ and $V(t)$ in Eq. (1.16) are given. Assume that at $t = 0$, $v = v_0$, and the sand is being added to the cart for $t > 0$. Find $v(t)$ for the following cases:

   (a) $V(t)$ arbitrary, $m(t) = \alpha t$, where $\alpha =$const. Keep the answer in terms of the antiderivative of $V(t)$.
   (b) $V(t) = $ const, $m(t)$ arbitrary (but given).

3. Suppose a rocket has the beginning mass M before burning fuel and the end mass m after burning fuel. Suppose the exit speed of gases exiting the rocket, *measured with respect to the rocket*, is V. Prove Tsiolkovsky's formula connecting the end speed of the rocket in the lab frame V_f to V and the ratio of masses M/m. Use the linear conservation law. Neglect all friction and gravity forces.
   Note There are lots of solutions for this problem available on the Internet. If you do consult the Internet, please be aware of which coordinate frame is being used, as some of these solutions are not particularly precise.
   Estimate what percentage of the rocket has to be taken by fuel to (a) reach the speed necessary to put a satellite in orbit and (b) to leave Earth's gravity forever. You may choose the exhaust velocity V either from modern rockets, from the dawn of the space era, or even ancient gunpowder rockets and compare the results.
4. In a Western movie, a good guy shoots a bad guy with a pistol during a standoff in a saloon. From the force of the impact, the aforementioned bad guy bounces back 2 m, performing a spectacular crash through the saloon doors. Consider the bad guy to be shot in the center of mass at 1 m height, starting the motion horizontally. Estimate the speed acquired by the bad guy needed to fly 2 m backward while dropping 1 m and the momentum of the bad guy from that bullet impact.
   Also, comment on what you think should happen to the good guy during this action scene.

Assuming that all the momentum of the bad guy was acquired from the bullet's impact, provide the following estimates:

(a) Using the typical speed of a bullet from a pistol or a rifle (either a current model or a historical one), compute the needed mass of the bullet to provide that momentum. Compare with the mass of a typical bullet.

(b) Using a typical mass of a bullet, estimate the speed of the bullet necessary to provide that momentum. Compare with the speeds yielded by modern or historic firearms.

5. You are riding on a cart with your friend, moving along a horizontal line without friction with velocity V. The cart is massless, and your mass is the same as your friend's. At $t = 0$, your friend steps off. You are asked to find the cart's speed after your friend leaves the cart.

(a) What is the speed of the cart if your friend steps off "gently," i.e., without pushing forward or backward? In other words, your friend moves at a speed equal to the cart at the moment of departure.

(b) If your friend jumps off the cart with velocity U relative to the cart, what is the final velocity of the cart?

6. A cart of mass M is moving without friction along a straight line. A block of dry ice is lying on the cart and is evaporating into the atmosphere, changing its mass as $m(t)$. At time T, all the dry ice has evaporated. What is the speed of the cart for $t > T$, given that the velocity of the system at time $t = 0$ was V_0?
Hint. What do you choose for $V(t)$ in Eq. (1.15) for this problem?

Chapter 2
From Newton's Laws to Euler-Lagrange Equations

2.1 Euler-Lagrange Equations

In this section, we take an "atomistic" view of a mechanical system. Suppose we separate the system, such as a pendulum or a rod, into N material particles. N can be as large as you wish, even going to atoms; then the number of particles chosen will play no role in the final equations. We will mark every material point with index i as $\mathbf{x}_i$. Every object you see around you will have a large number of material points. What is more important, however, is the *minimum* number of coordinates needed to describe the system. For example, a two-dimensional pendulum can be described by one coordinate, namely, the angle with respect to the vertical. A spherical pendulum will need two angles to describe its position. Thus, for each system, we define the coordinates $\mathbf{q} = (q^1, \ldots q^n)$ essential for describing the coordinates of each particle of the system $\mathbf{x}_i$ completely $\mathbf{x}_i = \mathbf{x}_i(\mathbf{q}, t)$. The coordinates q^α, $\alpha = 1, \ldots, n$ are called *generalized coordinates of mechanical system*. The choice of generalized coordinates is not unique, but the number of these coordinates, defining the dimension of the system, is specific to each mechanical system.

The first and second time derivatives of the i-th particle $\mathbf{x}_i(\mathbf{q}, t)$ are computed as

$$\dot{\mathbf{x}}_i = \frac{\partial \mathbf{x}_i}{\partial q^\beta} \dot{q}^\beta + \frac{\partial \mathbf{x}_i}{\partial t} = \dot{\mathbf{x}}_i(\dot{\mathbf{q}}, \mathbf{q}, t),$$

$$\ddot{\mathbf{x}}_i = \frac{\partial \dot{\mathbf{x}}_i}{\partial \dot{q}^\beta} \ddot{q}^\beta + \frac{\partial \dot{\mathbf{x}}_i}{\partial q^\beta} \dot{q}^\beta + \frac{\partial \dot{\mathbf{x}}_i}{\partial t},$$

(2.1)

where we have used the notation of *summation of repeated indices*. I shall use that notation throughout the book; if there is no sum over repeated indices, it will be indicated explicitly. Thus, there is the sum over β in the first and second equations of Eq. (2.1).

© The Author(s), under exclusive license to Springer Nature Switzerland AG 2025
V. Putkaradze, *A Concise Introduction to Classical Mechanics*,
Surveys and Tutorials in the Applied Mathematical Sciences 16,
https://doi.org/10.1007/978-3-031-84977-0_2

Suppose now that the forcing on the i-th particle can be separated into the potential (conservative), i.e., can be written in the form

$$\mathbf{F}_i = -\frac{\partial U}{\partial \mathbf{x}_i} + \mathbf{Q}_i, \tag{2.2}$$

where $U(\mathbf{x}^1, \mathbf{x}^2, \ldots, \mathbf{x}^n)$ is some function having the physical meaning of the potential for all particles, and $\mathbf{Q}_i$ is some kind of other nonconservative force acting on the i-th particle, for example, friction or external force applied to every particle. The second Newton's law states that

$$m_i \ddot{\mathbf{x}}_i + \frac{\partial U}{\partial \mathbf{x}_i} = \mathbf{Q}_i, \quad i = 1, \ldots, N. \tag{2.3}$$

However, Eq. (2.3) is not very useful. There may be an order of $N = 10^{23}$ particles (Avogadro's number) in a small rock, but if we made a 2D pendulum out of that rock, we would have only one degree of freedom (angle), and thus only one equation, not 10^{23} equations. So, Eq. (2.3) is dependent, and we need to project it on the coordinates q^α for every $\alpha = 1, \ldots, m$. The way to do it is to multiply Eq. (2.3) by a vector $\frac{\partial \mathbf{x}^i}{\partial q^\alpha}$ for every i and every α and then sum over all i. We also need to use $\ddot{\mathbf{x}}_i$ from Eq. (2.1). For every α, we have

$$
\begin{aligned}
0 &= \left(m_i \ddot{\mathbf{x}}_i + \frac{\partial U}{\partial \mathbf{x}_i} - \mathbf{Q}_i \right) \cdot \frac{\partial \mathbf{x}_i}{\partial q^\alpha} \\
&= \left(\frac{\partial \dot{\mathbf{x}}_i}{\partial \dot{q}^\beta} \ddot{q}^\beta + \frac{\partial \dot{\mathbf{x}}_i}{\partial q^\beta} \dot{q}^\beta + \frac{\partial \dot{\mathbf{x}}_i}{\partial t} + \frac{\partial U}{\partial \mathbf{x}_i} - \mathbf{Q}_i \right) \cdot \frac{\partial \mathbf{x}_i}{\partial q^\alpha}.
\end{aligned}
\tag{2.4}
$$

One important consequence of the formula for the time derivative $\dot{\mathbf{x}}_i$ given by Eq. (2.1) is the *cancellation of dots*. From the first equation of Eq. (2.1), we see by direct differentiation with respect to $\dot{q}^\alpha$ that

$$\frac{\partial \dot{\mathbf{x}}_i}{\partial \dot{q}^\alpha} = \frac{\partial \mathbf{x}_i}{\partial q^\alpha}. \tag{2.5}$$

We use that cancellation of dots formula to rewrite Eq. (2.4) as

$$\left(\frac{\partial \mathbf{x}_i}{\partial q^\beta} \ddot{q}^\beta + \frac{\partial \dot{\mathbf{x}}_i}{\partial q^\beta} \dot{q}^\beta + \frac{\partial \dot{\mathbf{x}}_i}{\partial t} + \frac{\partial U}{\partial \mathbf{x}^i} - \mathbf{Q}_i \right) \cdot \frac{\partial \mathbf{x}^i}{\partial q^\alpha} = 0. \tag{2.6}$$

Let us compute the kinetic energy of all the particles using Eq. (2.1) as

$$T(\mathbf{q}, \dot{\mathbf{q}}) = \frac{1}{2} m_i |\dot{\mathbf{x}}_i|^2 = \frac{1}{2} m_i \left(\frac{\partial \mathbf{x}_i}{\partial q^\alpha} \frac{\partial \mathbf{x}_i}{\partial q^\beta} \dot{q}^\alpha \dot{q}^\beta + 2 \frac{\partial \mathbf{x}_i}{\partial q^\alpha} \frac{\partial \mathbf{x}_i}{\partial t} \dot{q}^\alpha + \left| \frac{\partial \mathbf{x}_i}{\partial t} \right|^2 \right). \tag{2.7}$$

We can now notice a crucial result—the derivatives of $T(\mathbf{q}, \dot{\mathbf{q}})$ with respect to $\dot{q}^\alpha$ and time are directly connected with accelerations and thus Newton's law:

$$\frac{\partial T}{\partial \dot{q}^\alpha} = m_i \left(\frac{\partial \mathbf{x}_i}{\partial q^\beta} \dot{q}^\beta + \frac{\partial \mathbf{x}_i}{\partial t} \right) \cdot \frac{\partial \mathbf{x}_i}{\partial q^\alpha} = m_i \dot{\mathbf{x}}_i \cdot \frac{\partial \mathbf{x}_i}{\partial q^\alpha},$$

$$\frac{d}{dt} \frac{\partial T}{\partial \dot{q}^\alpha} = m_i \ddot{\mathbf{x}}_i \cdot \frac{\partial \mathbf{x}_i}{\partial q^\alpha} + m_i \dot{\mathbf{x}}_i \cdot \frac{\partial \dot{\mathbf{x}}_i}{\partial q^\alpha} = m_i \ddot{\mathbf{x}}_i \cdot \frac{\partial \mathbf{x}_i}{\partial q^\alpha} + \frac{\partial T}{\partial q^\alpha}, \qquad (2.8)$$

$$m_i \ddot{\mathbf{x}}_i \cdot \frac{\partial \mathbf{x}_i}{\partial q^\alpha} = \frac{d}{dt} \frac{\partial T}{\partial \dot{q}^\alpha} - \frac{\partial T}{\partial q^\alpha}.$$

Then, the version of combined Newton's law "projected" on coordinate q^α in the sense of (2.6) is written as

$$\frac{d}{dt} \frac{\partial T}{\partial \dot{q}^\alpha} - \frac{\partial T}{\partial q^\alpha} + \frac{\partial V}{\partial q^\alpha} = Q_\alpha, \quad \text{where}$$

$$V(\mathbf{q}, t) := U(\mathbf{x}_1(\mathbf{q}, t), \ldots, \mathbf{x}_N(\mathbf{q}, t)), \quad Q_\alpha := \mathbf{Q}_i \cdot \frac{\partial \mathbf{x}_i}{\partial q^\alpha}. \qquad (2.9)$$

Since only the kinetic energy T is assumed to depend on the velocities $\dot{\mathbf{q}}$, we can introduce the function called the *Lagrangian*, which is the difference between the kinetic and the potential energy:

$$L := T(\mathbf{q}, \dot{\mathbf{q}}) - V(\mathbf{q}) = \frac{1}{2} m_i |\dot{\mathbf{x}}_i(\mathbf{q})|^2 - V(\mathbf{q}) = \frac{1}{2} m_i \frac{\partial \mathbf{x}_i}{\partial q^\beta} \dot{q}^\beta \cdot \frac{\partial \mathbf{x}_i}{\partial q^\gamma} \dot{q}^\gamma - V(\mathbf{q})$$

$$(2.10)$$

(remember that the expression for the Lagrangian (Eq. (2.10)) includes the summation over the repeated indices i, β and γ). Then, Eq. (2.9) can be written in terms of the Lagrangian L only as

$$\frac{d}{dt} \frac{\partial L}{\partial \dot{q}^\alpha} - \frac{\partial L}{\partial q^\alpha} = Q_\alpha, \quad \alpha = 1, \ldots n, \quad L = T - V. \qquad (2.11)$$

These are the celebrated Euler-Lagrange equations for a mechanical system. They are ordinary differential equations that are second order in time with respect to coordinates q^α. To find a solution to these equations for $t > t_0$, we will need to specify a time $q^\alpha(t_0)$ and $\dot{q}^\alpha(t_0)$, i.e., the initial positions and generalized velocities.

Notice how all the subdivisions into particles $\mathbf{x}_i$ in Eq. (2.11) have disappeared. The only information left is given in terms of the essential coordinates of the system, involving time derivatives and derivatives with respect to q^α and $\dot{q}^\alpha$.

The steps for writing equations of motion for any mechanical system are as follows:

1. Find the "essential coordinates" q^α for describing the mechanical system. The number of these coordinates is the dimension of the system n, $\mathbf{q} = (q^1, \ldots, q^n)$.

2. Compute the Lagrangian $L(\mathbf{q}, \dot{\mathbf{q}})$ by calculating the kinetic $T(\mathbf{q}, \dot{\mathbf{q}})$ and potential energy $V(\mathbf{q})$ and defining $L = T - V$.
3. If appropriate, compute external nonconservative forces Q_α.
4. For each $\alpha = 1, \ldots, n$, compute the derivatives of L with respect to q^α and $\dot{q}^\alpha$ and write the Euler-Lagrange equations (Eq. (2.11)).
5. Depending on the problem, either solve a particular initial value problem for Eq. (2.11) or analyze the behavior of solutions for different values of parameters.

Digression: Cyclic Variables Assume there are no nonconservative forces, so $Q_\alpha = 0$ for $\alpha = 1, \ldots, n$. Suppose that for some i, Lagrangian is independent of the variable q^i, although it will, of course, depend on the variable $\dot{q}_i$. In that case, taking $\alpha = i$, and putting $\frac{\partial L}{\partial q^i} = 0$ in Eq. (2.11), we get

$$\frac{d}{dt} \frac{\partial L}{\partial \dot{q}^i} = 0 \quad \Rightarrow \quad \frac{\partial L}{\partial \dot{q}^i} = M_i = \text{const.} \tag{2.12}$$

This result is known as the conservation of generalized momentum for a cyclic variable. Trivial as this result may seem right now, it is actually very deep and important, which we will discover later in the framework of Noether's theorem in Chap. 4. More immediately, we will see the application of this result for the motion of planets in Kepler's law in Sect. 2.3 immediately below.

In Chap. 3, we shall quantify what we actually mean by the somewhat loose definition of "essential coordinates." The right mathematical term is the *configuration manifold* and its *tangent bundle*. For now, we shall talk about energy conservation, defined appropriately, as one of the most important consequences of the Euler-Lagrange equations (Eq. (2.11)).

2.2 Energy Definition and Conservation

Again, assume that there are no external forces, so $Q_\alpha = 0$ in Eq. (2.11). Given a Lagrangian $L = L(\mathbf{q}, \dot{\mathbf{q}}, t)$, with $\mathbf{q} = (q^1, \ldots, q^n)$, let us define a quantity E, which we call *energy*:

$$E = \dot{q}^\alpha \frac{\partial L}{\partial \dot{q}^\alpha} - L \quad \text{(sum over } \alpha\text{)}. \tag{2.13}$$

Let us compute $\dot{E}$ as follows:

$$\dot{E} = \ddot{q}^\alpha \frac{\partial L}{\partial \dot{q}^\alpha} + \dot{q}^\alpha \frac{d}{dt} \frac{\partial L}{\partial \dot{q}^\alpha} - \frac{\partial L}{\partial q^\alpha} \dot{q}^\alpha - \frac{\partial L}{\partial q^\alpha} \dot{q}^\alpha - \frac{\partial L}{\partial t}$$

$$= \dot{q}^\alpha \underbrace{\left(\frac{d}{dt} \frac{\partial L}{\partial \dot{q}^\alpha} - \frac{\partial L}{\partial q^\alpha} \right)}_{=0 \text{ by E-L eqs}} - \frac{\partial L}{\partial t} = -\frac{\partial L}{\partial t}. \tag{2.14}$$

Thus, if the Lagrangian is time-independent, then the quantity E we have called energy is (Eq. (2.13)) conserved, $\dot{E} = 0$.

Let us now clarify the physical meaning of this quantity E. Let us come back to the "atomistic" description of the system and consider the coordinate of each particle $\mathbf{x}_i \in \mathbb{R}^3$, $i = 1, \ldots, N$. Because $\mathbf{q}$ are the "essential" coordinates of the system, we must have $\mathbf{x}_i = \mathbf{x}_i(\mathbf{q}, t)$.[1] The kinetic energy T is computed as (sum over all repeated indices)

$$T = \frac{1}{2} m_i |\dot{\mathbf{x}}_i|^2 = \frac{1}{2} m_i \left| \frac{\partial \mathbf{x}_i}{\partial q^\alpha} \dot{q}^\alpha + \frac{\partial \mathbf{x}_i}{\partial t} \right|^2 = \frac{1}{2} G_{\alpha\beta}(\mathbf{q}, t) \dot{q}^\alpha \dot{q}^\beta + B_\alpha(\mathbf{q}, t) \dot{q}^\alpha + C(\mathbf{q}, t),$$
(2.15)

where $G_{\alpha\beta}$, B_α, and C are the functions of $\mathbf{q}$ and t, which can be computed in terms of the derivatives of $\mathbf{x}_i$ (sum over repeated index i):

$$G_{\alpha\beta} = m_i \frac{\partial \mathbf{x}_i}{\partial q^\alpha} \cdot \frac{\partial \mathbf{x}_i}{\partial q^\beta}, \quad B_\alpha = m_i \frac{\partial \mathbf{x}_i}{\partial q^\alpha} \cdot \frac{\partial \mathbf{x}_i}{\partial t}, \quad C = \frac{1}{2} m_i \frac{\partial \mathbf{x}_i}{\partial t} \cdot \frac{\partial \mathbf{x}_i}{\partial t}. \quad (2.16)$$

In general, one can show that $G_{\alpha\beta}$ should be nondegenerate for particle systems. Note that G is symmetric: $G_{\alpha\beta} = G_{\beta\alpha}$. The important feature of Eq. (2.15) that kinetic energy defined by Eq. (2.15) is a quadratic function of generalized velocities $\dot{q}^\alpha$. With this expression for the kinetic energy, the energy E defined in Eq. (2.13) is given by

$$E = \dot{q}^\alpha \left(B_\alpha + G_{\alpha\beta} \dot{q}^\beta \right) - \left(\frac{1}{2} G_{\alpha\beta} \dot{q}^\alpha \dot{q}^\beta + B_\alpha \dot{q}^\alpha + C - V \right)$$
$$= \frac{1}{2} G_{\alpha\beta}(\mathbf{q}, t) \dot{q}^\alpha \dot{q}^\beta + V - C.$$
(2.17)

Thus, *if* the coordinates $\mathbf{x}^i$ are independent of time, then $B_\alpha = 0$ and $C = 0$ in Eq. (2.16), so the energy defined by Eq. (2.13) is equal to just kinetic plus potential energy. Otherwise, the quantity (Eq. (2.13)) is not equal to the commonly understood concept of total energy and differs by the quantity C defined in Eq. (2.16). This happens when the coordinates depend explicitly on time. We will see examples of the choice of such coordinates for the case of particles moving on surfaces. An example of such cases is the motion of a bead on a hoop, rotating with a fixed angular velocity, solved in Sect. 3.7.

---

[1] We made $\mathbf{x}_i$ depend on both $\mathbf{q}$ and t for generality. In reality, we will most often have $\mathbf{x}_i$ depend only on $\mathbf{q}$. We shall see the difference that the time dependence of $\mathbf{x}_i(\mathbf{q}, t)$ introduces below.

2.3 Motion in Central Potential and Kepler's Law

As an example of the power of Euler-Lagrange equations, we will now solve perhaps the most celebrated problem of all times—Newton's results justifying Kepler's laws of planetary motion. This derivation describes the motion of just one particle, a planet, rotating around the Sun. Our solution is, of course, not the one Newton has found: he used his laws of motion to solve the problem. However, this problem of planetary motion can also be solved using the Euler-Lagrange equations, which is an illuminating example of how useful and elegant solutions based on Euler-Lagrange equations can be.

In 1609, Johannes Kepler published the celebrated laws of planetary motion—built purely on the studies of planetary motion from observation:

1. The orbit of a planet is an ellipse with the Sun at one of the two foci.
2. A line segment joining a planet and the Sun sweeps out equal areas during equal intervals of time.
3. The square of a planet's orbital period is proportional to the cube of the length of the major semiaxis of its orbit.

The mathematical explanation of the origin of these laws was done in an incredible tour de force by Newton. We will now explain these laws using the Euler-Lagrange equations. It is, of course, not the path Newton chose since Euler-Lagrange equations didn't exist at his time. The geometry of the problem and chosen coordinate systems are presented in Fig. 2.1.

Let us use the Euler-Lagrange equations (Eq. (2.11)) to describe the motion of two particles moving in a gravitational field. For example, it could be a Sun and a planet moving around the common center of mass. Since we assume that the system

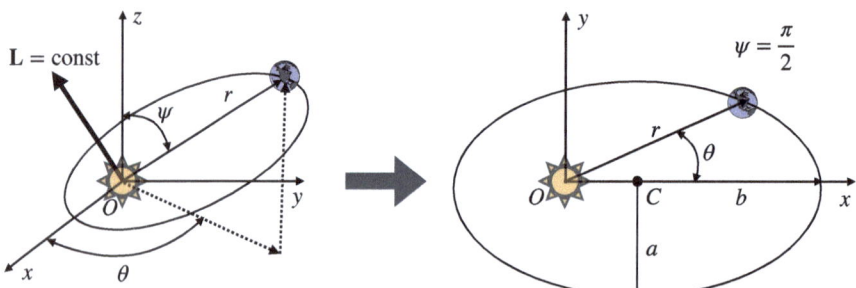

Fig. 2.1 The geometry of Kepler's problem. Left panel: the planet is moving around the Sun on some trajectory. The Sun is at the center of the coordinate system (point O). We choose the spherical coordinate system (r, θ, ψ). We then notice that because of the conservation of the angular momentum $\mathbf{L}$, the trajectory has to be planar with $\psi = \pi/2$ and can be described in coordinates (r, θ) as illustrated on the right panel. The center of the ellipse, which, as Newton has proved, is the trajectory of the planet, is at the point C, and a and b are the major semiaxes of the ellipse, with the Sun at one of the foci of the ellipse

is isolated, we can choose a coordinate system for which the center of mass is at the origin of $\mathbb{R}^3$, i.e., $r_{CM} = 0$. Each particle is moving, so the position of the center of mass remains in 0 for all times. Thus, it is sufficient to compute the motion of one particle only since the motion of the other particle can be computed simply by extending the center of mass by an appropriate distance beyond the origin. Thus, we shall only consider the evolution of one particle with the coordinate $r \in \mathbb{R}^3$ about the origin.

We shall only be interested in the case of the central potential, i.e., a potential depending only on the distance from the origin $r = |r|$ only: $V(r) = V_0(r)$.[2] This assumption of central potential means that the force acting on the particle $\mathbf{F} = -\nabla V = -V_0'(r)r/|r|$ is directed toward the origin. This is the particular physical case obeyed by the motion of planets in the gravitational field of a star, also known as Kepler's problem.

In that case, $V_0 = -mMG/r$, where G is the gravitational constant and m and M are masses of the planet and the star, respectively. In what follows, we will combine the mass of the Sun M and gravitational constant G into one constant, k, and write the force acting on the planet of mass m as

$$V = -\frac{km}{r}, \quad \mathbf{F} = -\nabla V = -km\frac{r}{r^3}, \quad r := |r|. \tag{2.18}$$

The units of k are length^3/time^2. There are important consequences of equations that follow for an arbitrary central potential function $V_0(r)$, which we will explore first.

Let us compute the motion of a planet in spherical coordinates $r = r(\sin\psi\cos\theta, \sin\psi\sin\theta, \cos\psi)$ ($\psi = 0$ corresponds to the vertical axis). We shall use (r, ψ, θ) as the variables $\mathbf{q}$ in Eq. (2.11), so $q^1 = r$, $q^2 = \psi$, $q^3 = \theta$. Then, the kinetic energy is given by

$$T = \frac{1}{2}m|\dot{r}|^2 = \frac{1}{2}m\left(\dot{r}^2 + r^2\dot{\psi}^2 + r^2\dot{\theta}^2\sin^2\psi\right). \tag{2.19}$$

The Lagrangian is thus

$$L = T - V_0(r) = \frac{1}{2}m|\dot{r}|^2 - V(r) = \frac{1}{2}m\left(\dot{r}^2 + r^2\dot{\psi}^2 + r^2\dot{\theta}^2\sin^2\psi\right) - V_0(r). \tag{2.20}$$

We shall note that an equivalent system is obtained when computing the motion of two isolated particles moving about the common center of mass and interacting through the potential depending only on the distance between the particles.

---

[2] Here, $r = (r_1, r_2, r_3)$ is a vector in a 3D space, and $|r| = \sqrt{r_1^2 + r_2^2 + r_3^2}$.

Let us compute the number of conserved quantities we can find.

1. Since the Lagrangian (Eq. (2.20)) does not depend on time, the energy $E = T + V_0(r)$ is conserved.
2. Since the force $F = -\nabla V$ is parallel to the vector pointing to the origin, the net torque $\mathbf{T}$ in Eq. (1.13) vanishes, so the angular momentum $\mathbf{L}$ is conserved. That gives, nominally, three conserved quantities.

Remark 2.1 (Additional Constants of Motion for Kepler's Problem) For the gravitational potential $V_0 = -km/r$, there are other conserved quantities, namely, the Laplace-Runge-Lenz vector:

$$\mathbf{A} = \mathbf{p} \times \mathbf{L} - mk\frac{\mathbf{r}}{r}, \qquad (2.21)$$

with $\mathbf{p}$ being the momentum of the particle. The Euler-Lagrange equations for Kepler's problem have dimension six since the equations are of second order in the three-dimensional space variables r. A system of dimension 6 can have at most five independent constants of motion to show any kind of nontrivial dynamics. Thus, seven constants of motion E, $\mathbf{L}$, and $\mathbf{A}$ cannot be independent. There are just five independent constants of motion to make this problem *super-integrable*: these five constants of motion define a curve in the six-dimensional phase space. Kepler's problem is also known by the name maximally super-integrable since there are exactly five independent constants of motion.

The Motion in a Central Potential Occurs in a Plane As a preparation to demonstrate the first of Kepler's laws, we will prove that the motion occurs in the plane. Let us take the plane normal to the angular momentum vector $\mathbf{L}$, which is a constant of motion. First, let us notice that all the motion has to be in that plane.

Indeed, take an arbitrary point in time. The vector r is in the plane normal to the fixed vector $\mathbf{L} = r \times m\dot{r}$ (we say that $\mathbf{L}$ is fixed since $\mathbf{L}$ is conserved). Then, if $\dot{r}$ had a component normal to that plane (parallel to $\mathbf{L}$), then $\mathbf{L} = r \times m\dot{r}$ will have a component parallel to the plane, and thus normal to itself, which is impossible. Thus, we can assume that the motion of the system is planar.

Let us choose the center of mass in that plane. Next, we verify that the motion in the plane is indeed the solution for the Euler-Lagrange equations. For the ψ coordinate, we get

$$\frac{d}{dt}\frac{\partial L}{\partial \dot{\psi}} - \frac{\partial L}{\partial \psi} = 0, \quad \Rightarrow \quad mr^2\left(\ddot{\psi} - \sin\psi\cos\psi\dot{\theta}^2\right) = 0. \qquad (2.22)$$

Clearly, $\psi = 0$ or $\psi = \pi/2$ is a solution of Eq. (2.22), corresponding to the point moving either on the axis of $\mathbf{L}$ or in the plane normal to $\mathbf{L}$ containing the position of the center of mass. Even though $\dot{\theta}$ enters Eq. (2.22), the solutions $\psi = 0$ and $\psi = \pi/2$ are valid no matter what θ is, as the term $\dot{\theta}^2$ is multiplied by 0. We choose the second option $\psi = \pi/2$, so the motion is along the plane normal to the axis. This argument, showing the reduction from the three-dimensional to planar motion, is shown in Fig. 2.1, from the left panel, illustrating the 3D motion, to the right panel, showing the reduction for the 2D motion and $\psi = \pi/2$.

Kepler's Second Law In the plane $\psi = \pi/2$, the motion is reduced to (r, θ) coordinates with the Lagrangian:

$$L = \frac{1}{2}m(\dot{r}^2 + r^2\dot{\theta}^2) - V_0(r). \tag{2.23}$$

Since the Lagrangian does not depend on the coordinate θ, the Euler-Lagrange equation for θ can be integrated:

$$\frac{d}{dt}mr^2\dot{\theta} = 0 \quad \Rightarrow \quad l = mr^2\dot{\theta} = \text{const}. \tag{2.24}$$

Note that $\dot{A} = \frac{1}{2}r^2\dot{\theta}$ is precisely the quantity described in Kepler's second law, namely, the area swept by a vector from the Sun to the planet per (infinitesimal) unit of time, as can be seen from the right panel of Fig. 2.1. Thus, the whole area of the trajectory A is swept by the vector in time:

$$T = \frac{A}{\dot{A}} = \frac{2Am}{l}, \tag{2.25}$$

which is the period of the trajectory. The final equation for the r component only is obtained by substituting $\dot{\theta}$ from Eq. (2.24):

$$\frac{d}{dt}m\dot{r} - mr\dot{\theta}^2 + \frac{dV_0}{dr} = 0 \quad \Rightarrow \quad m\ddot{r} - \frac{l^2}{mr^3} + \frac{dV_0}{dr} = 0. \tag{2.26}$$

We can define the effective potential such that

$$m\ddot{r} = -W'(r), \quad W(r) = \frac{l^2}{2mr^2} + V_0(r). \tag{2.27}$$

Thus, the motion of any particle in a central potential reduces to a one-dimensional problem in variable r. The effective energy of motion computed for r variables only, $E = \frac{1}{2}m\dot{r}^2 + W(r)$, is conserved. We can use that conservation of energy to compute phase portraits in the $(r, \dot{r})$ plane by choosing a curve $E = E_0 = \text{const}$. Below, we present some examples before tackling Kepler's problem, which can be solved analytically.

Motion in Elastic Potential $V_0(r) = Kr^2/2$ This is the case when the planet is connected with the Sun by an elastic spring. The effective potential for a fixed l and the energy are

$$W_l = \frac{l^2}{2mr^2} + \frac{1}{2}Kr^2, \quad E_l = \frac{1}{2}m\dot{r}^2 + \frac{l^2}{2mr^2} + \frac{1}{2}Kr^2 = \text{const}. \tag{2.28}$$

In order to plot solutions in $(r, \dot{r})$ plane, we choose $E_l = C$ to be a fixed constant and select the intersection of $E = E_l = C$ with the surface E_l defined by (2.28).

Kepler's Laws Suppose $V_0 = -mk/r$. The equations of motion (Eq. (2.26)) are

$$\ddot{r} - \frac{l^2}{m^2 r^3} + \frac{k}{r^2} = 0 . \tag{2.29}$$

The key to the solution is to take $r = r(\theta)$. Then,

$$\dot{r} = \dot{\theta}\frac{dr}{d\theta} = \frac{l}{mr^2}\frac{dr}{d\theta}, \quad \ddot{r} = \frac{l}{mr^2}\frac{d}{d\theta}\left(\frac{l}{mr^2}\frac{dr}{d\theta}\right). \tag{2.30}$$

The key to the solution is to notice that the substitution $r = 1/u(\theta)$ is useful since it simplifies the expression in parentheses on the right-hand side of Eq. (2.30). Using that substitution, and multiplying Eq. (2.29) by mr^2/l^2, we obtain a linear, second-order equation for $u(\theta)$:

$$\frac{d^2u}{d\theta^2} + u = \frac{m^2k}{l^2} \tag{2.31}$$

The solution of Eq. (2.31) is readily found as

$$u = \frac{1}{r} = \frac{m^2k}{l^2} - \alpha\cos(\theta - \theta_0) , \tag{2.32}$$

with α and θ_0 depending on the initial conditions. The type of the solution depends on the relationship between a and the constant term m^2k/l^2. There are three options:

1. If $\alpha < \frac{m^2k}{l^2}$, then $u > 0$ and Eq. (2.32) specifies an ellipse in the plane of the orbit.
2. If $\alpha = \frac{m^2k}{l^2}$, then $u \to 0$ as $\theta \to \theta_0$. The solution is a parabola.
3. If $\alpha > \frac{m^2k}{l^2}$, then $u \to 0$ as $\theta \to \theta_0 \pm \theta_* = \pm\arccos\frac{m^2k}{l^2\alpha}$. The trajectory of the particle is a hyperbola.

This completes the derivation of Kepler's first law. You can notice that we have had three possible answers for the shape of the trajectory: an ellipse, a hyperbola, or a parabola. The planets, of course, can only follow elliptical trajectories since their distance to the Sun is bounded.

Kepler's Third Law To prove Kepler's third law, it is more convenient to write the solution (Eq. (2.32)) in the standard form of an ellipse:

$$r = \frac{p}{1 - \epsilon\cos(\theta - \theta_0)}, \quad p := \frac{l^2}{m^2k}, \quad \epsilon = \frac{\alpha}{p}, \tag{2.33}$$

where, for an elliptical trajectory, the eccentricity is small: $\epsilon = a/p < 1$. The major a and minor b semiaxis and the area of this ellipse and its area A are given by

$$a = \frac{p}{1 - \epsilon^2}, \quad b = a\sqrt{1 - \epsilon^2}, \quad A = \pi ab = \pi a^2\sqrt{1 - \epsilon^2}. \tag{2.34}$$

We see from Eq. (2.33) that $l^2 = pm^2k = m^2ka(1 - \epsilon^2)$. To obtain Kepler's third law, from Eq. (2.25), we conclude that

$$T^2 = \frac{4A^2m^2}{l^2} = \frac{4\pi^2m^2a^4(1 - \epsilon^2)}{l^2} = \frac{4\pi^2}{k}a^3, \tag{2.35}$$

which is exactly Kepler's third law.

2.4 Looking Forward

The calculation by Newton in the famous *Principia Mathematica* [12], leading to the results presented above, jump-started the field of mechanics hundreds of years ago. Newton's *Principia* is, no doubt, very hard to read for a modern reader, especially as one of your first books in mechanics, should you choose to do so. You may want to read a beautiful retelling of the story by Chandrasekhar [13]. Does it mean that this whole discussion is ancient history?

Certainly not. Newton's solution to Kepler's problem has opened the whole field of integrable systems. For example, as it turns out, the reason that we can solve Schrödinger's equation for the hydrogen atom is connected with the super-integrability of Newton's problem [14]. For example, a lot of the attention of the scientific community nowadays has been focusing on the study of multi-body problems in celestial mechanics, in particular the motion of three planets (the three-body problem). For example, perturbation of Kepler's orbits of Uranus led to the discovery of Neptune in 1846 [15–18]. Similar methods are now helping with finding exoplanets [19]. There are now books [20] and review articles [21] dedicated to the three-point problem and beautiful exact solutions found recently, formed by "dancing" closed solutions—aptly called "choreographies" [22]. In spite of over 300 years of history, planetary motion and celestial mechanics is still a very active area of research, especially concerning applications to astronomy, astrophysics, and space travel.

2.5 Practice Problems

1. Consider the motion of a particle in the central potential $U(r) = Kr^p$, $p > 0$. Find an effective potential (as in the notes) and sketch the trajectories on $(r, \dot{r})$ plane.

2. Consider the motion of a particle in the central potential $U(r) = -Kr^{-p}$, $p > 0$. Find an effective potential (as in the notes) and sketch the trajectories on $(r, \dot{r})$ plane. Consider the cases $0 < p < 2$, $p = 2$, and $p > 2$.

3. Suppose the Earth is stopped in its motion in space and starts falling on the Sun. How much time will the fall take? Obtain the answer without computing any integrals or solving any differential equations.
   Hint Use Kepler's third law.

4. Consider a 1D motion of a particle on a line, $q \in \mathbb{R}$, and consider the Lagrangian $L = \dot{q} - q$. Write the Euler-Lagrange equations for this system. What happened?

Chapter 3
Configuration Manifolds, Variational Principle, and Euler-Lagrange Equations

3.1 Preliminaries: Notation and Definitions

In what follows, and everywhere in this book, we assume that the material particles are moving in the familiar three-dimensional space, which is denoted $\mathbb{R}^3$. Vectors in $\mathbb{R}^3$ are always denoted with a bold case text and are defined as the column vectors:

$$\mathbf{x} = \begin{pmatrix} x^1 \\ x^2 \\ x^3 \end{pmatrix} = \text{(a column)}. \tag{3.1}$$

Why do we have an index on the top of x's? That is, to distinguish the vectors from the so-called covectors, which are going to be denoted with indices down and correspond to the row vectors:

$$\mathbf{f} = \begin{pmatrix} f_1 & f_2 & f_3 \end{pmatrix} = \text{(a row)}. \tag{3.2}$$

A trajectory in space is a curve $\mathbf{x}(t)$, where t is time, taken as a scalar parameter. The velocity $\mathbf{v}$ is the time derivative of $\mathbf{x}(t)$, which we denote with a dot above $\mathbf{x}$, i.e., $\mathbf{v} = \dot{\mathbf{x}}$. The acceleration $\mathbf{a}$ is the time derivative of $\mathbf{v}$, i.e., $\mathbf{a} = \dot{\mathbf{v}} = \ddot{\mathbf{x}}$. Using this simple argument, we arrive at the conclusion that *velocity is a vector*: $v^i = \dot{x}^i$. What is a familiar example of a covector? For example, the gradient of a scalar is a covector:

$$\mathbf{f} = \frac{\partial U(\mathbf{x})}{\partial \mathbf{x}} \quad \Leftrightarrow \quad f_i = \frac{\partial U}{\partial x^i}. \tag{3.3}$$

Notice that the position of indices for $\mathbf{f}$ (down) follows the position of indices in the derivative.

© The Author(s), under exclusive license to Springer Nature Switzerland AG 2025
V. Putkaradze, *A Concise Introduction to Classical Mechanics*,
Surveys and Tutorials in the Applied Mathematical Sciences 16,
https://doi.org/10.1007/978-3-031-84977-0_3

Why are we so insistent on distinguishing the indices as being up or down? It is useful to acquire good habits of keeping the indices in the proper places. When working in the Euclidian 3D space only, it may be acceptable, since formally, "vectors look the same as covectors".[1] However, this simple relationship will break when going to, for example, non-Cartesian coordinate systems or performing computations on the surfaces.

We introduce the following:

Definition 3.1 We shall call the quantities with indices up *contravariant* and indices down *covariant* variables.

Take the simplest case of $\mathbb{R}^n$, and suppose $B = \{\mathbf{e}_1, \ldots, \mathbf{e}_n\}$ is a basis in that space. If you take a vector $\mathbf{x} = x^i \mathbf{e}_i$ (summation over repeated indices) and change the basis to another basis, say $\widetilde{B} = \{\widetilde{\mathbf{e}}_1, \ldots, \widetilde{\mathbf{e}}_n\}$, then while the "physical" vector $\mathbf{x}$ remains the same, the coordinates of $\mathbf{x}$ in the new basis will be different:

$$\mathbf{x} = \sum_{i=1}^{n} x^i \mathbf{e}_i = \sum_{i=1}^{n} \widetilde{x}^i \, \widetilde{\mathbf{e}}_i \, . \tag{3.4}$$

If the vectors of the old and new basis in Eq. (3.4) are connected through $\widetilde{\mathbf{e}}_j = M^i_{\ j} \mathbf{e}_i$, then, for consistency, we need to put $\widetilde{x}^j M^i_{\ j} = x^i$. In other words, the new and old *components* of the vector $\mathbf{x}$ are connected through the inverse of the transformation matrix $\mathbb{M}$ between the old and new components. Similarly, one can derive that the components of the covariant tensor change according to $\mathbb{M}$. In any case, the transformation of the coordinates is different for contra- and covariant vectors, and thus, physical laws equating covariant quantities with contravariant quantities cannot exist.

Even if you never intend to go to the non-Cartesian coordinate systems, equalling vectors and covectors is fraught with danger. Consider, for example, the following interesting application. Let us make a model of a particle moving in a viscous media with the law "velocity $\mathbf{v}$ is proportional to force $\mathbf{F}$"—which is a popular model for such physical phenomena. One is tempted to write

$$\mathbf{v} = k\mathbf{F} \, , \tag{3.5}$$

where k is some scalar and, for example, the force $\mathbf{F} = -\frac{\partial U(\mathbf{x})}{\partial \mathbf{x}}$ is coming from some potential. You immediately see that Eq. (3.5), as it stands, makes little mathematical sense: the velocity has indices up, and the force has indices down. Then, under coordinate transformations, $\mathbf{v}$ and $\mathbf{F}$ transform differently. For example, take Eq. (3.5) as it stands and ask your friends to look at the same law from their point of view. Your friends' coordinate frame is not identical to yours, and

---

[1] This statement is somewhat mathematically loose but could be formalized in terms of the metric tensor being unity.

thus, their coordinates $\mathbf{x}_f$ are different from yours. The velocity on the left-hand side transforms as a contravariant tensor, but the force $\mathbf{F}$ on the right-hand side transforms differently (as a covariant tensor), and thus the physical law (Eq. (3.5)), as it stands, would depend on the right choice of coordinates.

What was the error of a perfectly reasonably looking Eq. (3.5)? One should remember that any connection between vectors and covectors must involve lowering and raising indices. For example, we can make the equation consistent by adding a metric tensor on the left-hand side of Eq. (3.5). This example should, hopefully, provide some words of caution about writing laws and balancing quantities with different geometric types.

Now, let us study operations on vectors and covectors. We can naturally make a multiplication of a row and a vector, which we call a pairing:

$$\langle \mathbf{f}, \mathbf{x} \rangle := \sum_{i=1}^{3} f_i x^i . \tag{3.6}$$

We remind the reader that we normally drop the sum and use the summation over repeated indices (also known as Einstein's notation), so we just write

$$\langle \mathbf{f}, \mathbf{x} \rangle := f_i x^i . \tag{3.7}$$

The row vectors are often called *covectors*, and that is the notation we shall use. However, one can see that we will have to work a bit to make a product between two vectors $\mathbf{a}$ and $\mathbf{b}$ since one cannot simply multiply two column vectors. We proceed by defining the scalar product: take a symmetric, positive definite matrix $\mathbb{K}$ with two indices down, i.e., $K_{ij} = K_{ji}$, which also needs to satisfy some inequality constraint, and define

$$\mathbf{a} \cdot \mathbf{b} = K_{ij} a^i b^j . \tag{3.8}$$

Note that Eq. (3.8) can be written as $\langle \mathbb{K}\mathbf{a}, \mathbf{b} \rangle$, and so $\mathbb{K}\mathbf{a}$ makes a covector, which can then be paired with the vector $\mathbf{b}$. For Eq. (3.8) to make sense as a scalar product, we need to enforce $|\mathbf{a}|^2 \geq 0$, with the equality achieved only when $\mathbf{a} = \mathbf{0}$. This equation can be written as

$$(\mathbb{K}\mathbf{a}) \cdot \mathbf{a} \geq 0, \quad \text{and} \quad (\mathbb{K}\mathbf{a}) \cdot \mathbf{a} = 0 \quad \text{if and only if} \quad \mathbf{a} = \mathbf{0} . \tag{3.9}$$

For example, when $\mathbf{K} = \text{diag}(K_1, K_2, K_3)$, and $K_i > 0$, then Eq. (3.9) is satisfied (prove it!).

3.2   Manifolds and Tangent Bundles

Manifolds Loosely speaking, an n-dimensional manifold M is an object that can be represented, locally, by a set of coordinate charts $\{U_i\}_{i=1}^N$. These coordinate charts define local coordinates on $U_i \in M$ or, alternatively, introduce a mapping $\mathbb{R}^n \to U$. When the domains U_i and U_j have a nonzero intersection, there are at least two coordinates on $U_i \cup U_j$. These coordinates must be compatible; that is, there must be a sufficiently smooth mapping between them. A more precise definition is given as follows:

Definition 3.2 A smooth manifold M of dimension n is a set with the following properties:

1. There exists a finite, or perhaps a countable, set of domains U_i with the union of these sets being M: $\cup_i U_i = M$.
2. On each of these sets U_i, there is a one-to-one smooth mapping φ_i from U_i to a compact domain $V_i \subset \mathbb{R}^n$.
3. When the sets U_i and U_j have a non-empty intersection, then we can define a mapping from $U_i \cap U_j$ to itself using first φ_i mapping it into a set of $\mathbb{R}^n$, and then back using φ_j^{-1}. That composition of maps $\varphi_j^{-1}(\varphi_i),^2$ must be smooth.

Why, an acute reader may ask, do we need such a complex structure for mechanics? Didn't we hear about the phase space in our previous courses? The problem is that the concept of space is inadequate for describing most mechanical systems. It is only relevant for a set of N interacting particles in space—and little else. The "phase space" is not a space even for a pendulum, to say nothing about more complex systems![3]

Don't be scared by a somewhat complex definition (Eq. (3.2)). As we will see, any mechanical system we will encounter will have a configuration manifold. Indeed, most of the examples in mechanics we will encounter come from some restrictions (constraints) on the motion of some or all parts of a mechanical system. For example, the motion of a pendulum's bob on a plane happens on a circle since there is a restriction that the length of the pendulum l is fixed, which can be written as $|\mathbf{x}|^2 - l^2 = 0$. The following theorem, based on the implicit function theorem, guarantees that physically relevant restrictions of the type we encounter in mechanics lead to a manifold.

We consider a set in $\mathbf{x} \in \mathbb{R}^{n+m}$ that is given by a level set of functions $f_k(\mathbf{x}) = 0$, $k = 1, \ldots m$, a set that we denote $\mathbf{f}(\mathbf{x}) = \mathbf{0}$.

---

[2] also known as $\varphi_j^{-1} \circ \varphi_i$.

[3] Unless, of course, we do not want "space" to be a vector space, meaning essentially a manifold. But in that case, it is better to use the right vocabulary from the very beginning.

Definition 3.3 Point **0** is a regular value of the function $\mathbf{f}(\mathbf{x})$ if the gradient $\frac{\partial \mathbf{f}}{\partial \mathbf{x}}(\mathbf{x})$ (a matrix) is a surjective map between $\mathbb{R}^{n+m}$ and $\mathbb{R}^m$.[4]

Sometimes the set $\mathbf{f}(\mathbf{x}) = \mathbf{0}$ in $\mathbb{R}^{n+m}$ is denoted as $Q = \mathbf{f}^{-1}(\mathbf{0})$ We use the following useful theorem, which is connected to the implicit function theorem:

Theorem 3.4 *Suppose $\mathbf{f}(\mathbf{x}) = \mathbf{0}$ is a differentiable map between $\mathbb{R}^{n+m}$ and $\mathbb{R}^m$. If $\mathbf{0} \in \mathbb{R}^m$ is a regular value of $\mathbf{f}(\mathbf{x})$, then $Q = \mathbf{f}^{-1}(\mathbf{0})$ is a manifold of dimension n.*

In other words, imposing several reasonable physical constraints (i.e., with nonvanishing gradients) on a mechanical system always leads to the "essential coordinates" of the systems on a manifold.

It isn't easy to create an object that is not manifold, which is relevant for mechanics. There are, of course, sets that are not manifolds; it is just that they are hardly ever relevant for mechanics applications. One example of a set that is not a manifold is a Cantor set, which is relevant to the chaotic dynamics. A more familiar set that will give you some difficulty in the direct application of a manifold definition is figure eight or any line with self-intersections.

Let us look at some examples of sets that are manifolds:

1. *A circle* S^1 : $x^2 + y^2 = 1$. Consider the intervals $I_1 = (\pi/4, 7\pi/4)$ and $I_2 = (-\pi/4, 5\pi/4)$. You can parameterize S^1 by taking $\theta_i \in I_1$ and considering the maps $\phi_i : I_1 \rightarrow (\cos\theta_i, \sin\theta_i)$. The maps in the overlaps are just shifts in angle and are, of course, differentiable as many times as you wish. Thus, S^1 is a manifold.

   Alternatively, you can say that a circle is the level set $f = 0$ for the function $f(x, y) = x^2 + y^2 - 1$, $f : \mathbb{R}^2 \rightarrow \mathbb{R}$. Then, $\nabla f = (2x, 2y)$, and $|\nabla f|^2 = 4(x^2 + y^2) = 4 \neq 0$ when $x^2 + y^2 = 1$. Thus, 0 is a regular value for $f(x, y)$, and so the level set $f(x, y) = x^2 + y^2 - 1 = 0$ is a manifold.

2. *A two-dimensional sphere* S^2 : $x^2 + y^2 + z^2 = 1$. Take the sphere and remove a point (say, the North Pole). Then, take a stereographic projection of the sphere on the plane tangent to the sphere at the South Pole. This is the first coordinate chart φ_1 mapping the plane to the sphere without the North Pole. To get the second coordinate chart φ_2, repeat the procedure with the South Pole removed and use stereographic projection on the plane tangent at the North Pole. The mapping $\varphi_1^{-1} \circ \varphi_2$ is defined as a mapping from one plane to another and is clearly one-to-one and smooth. (Compute that mapping as an exercise!)

   Alternatively, similar to above, a sphere is the level set $f = 0$ for $f(x, y, z) = x^2 + y^2 + z^2 - 1$. Then, on that set, $|\nabla f|^2 = 4 \neq 0$. Thus, 0 is a regular value for $f(x, y, z)$, and hence, the level set of $f(x, y, z) = x^2 + y^2 + z^2 - 1 = 0$ is a manifold.

---

[4] A surjective function (onto function) is a function $F : X \rightarrow Y$ such that, for every element $y \in Y$, there exists at least one element $x \in X$ such that $F(x) = y$. In the case considered here, you can also say that the rank of the gradient matrix $\frac{\partial \mathbf{f}}{\partial \mathbf{x}}(\mathbf{x})$ is equal to m.

3. *A surface* Any surface given by a function $f(x_1, x_2)$, where $(x_1, x_2) \in V$ (an open, connected subset of $\mathbb{R}^2$), is a manifold.

3.3   Configuration Manifolds

We are now ready to define the configuration manifold of a mechanical system.

Definition 3.5 A set of variables $\mathbf{q} = (q^1, \ldots, q^n)$ uniquely defining a state of a mechanical system is called generalized coordinates. The set of all possible configurations of the system in generalized coordinates defines the configuration manifold.

Why do we say that the set of configurations is a manifold? It is often defined as a subset of a linear space with a finite number of algebraic constraints, which, according to Theorem 3.4, is a manifold. We shall often denote this manifold as Q. Let us proceed to some examples of configuration manifolds:

1. Pendulum swinging in a plane. The coordinates of the pendulum are uniquely defined by one angle. Thus, the pendulum has the configuration manifold $Q = S^1$.
2. Spherical pendulum. The position of a pendulum in space is uniquely determined by the director of a pendulum, which selects a point on a sphere. Thus, for a spherical pendulum, $Q = S^2$.
   Note. One can sometimes hear a statement like this. Since the direction of a spherical pendulum is defined by two angles, say θ and ϕ, and each angle is part of S^1, then the configuration manifold for a spherical pendulum is $Q = S^1 \times S^1 = T^2$, a two-dimensional torus. That clearly contradicts what we wrote above since a sphere is different from a torus. What goes wrong?
3. Double pendulum swinging in a plane. This system is described by two angles φ_1 and φ_2, and each of these angles can be selected anywhere on the circle. Thus, the configuration manifold is $Q = S^1 \times S^1 = T^2$, a two-dimensional torus.
4. A disk rolling inside a freely spinning circle. This system is described by the rotation of the circle and the position of the disk, for example, its rotation angle or its inclination with respect to the vertical. Thus, the configuration manifold is $Q = S^1 \times S^1 = T^2$, a two-dimensional torus.

The last two examples show that two very different mechanical systems can have the same configuration manifold. The manifold selects a way to describe the coordinates of the system $\mathbf{q}$. The physics is described by the Lagrangian function $L(\mathbf{q}, \dot{\mathbf{q}})$. However, before we can define the Lagrangian, we need to understand what is meant by the derivative $\dot{\mathbf{q}}$. To do so, we will need to introduce the concept of the tangent bundle.

3.4 Tangent Bundle of a Smooth Manifold

Suppose Q is an n-dimensional manifold. At each point of $q \in Q$, there is an open neighborhood U_q locally equivalent to an open subset $V \subset \mathbb{R}^n$. Suppose $\varphi(t)$ is the coordinate chart, and suppose y_0 is the pre-image of q: $\varphi(y_0) = q$. Take a smooth curve $y(t) \in U$, with $y(0) = y_0$, and $t \in [-\epsilon, \epsilon]$ for some $\epsilon > 0$. This curve creates a curve $q(t) = \varphi(y(t)) \in Q$. Thus, the mapping $\varphi(y)$ creates a map between the curves on the coordinate charts and on the manifold. Taking the derivatives of $y(t)$ at $t = 0$, the mapping $\varphi(t)$ also defines a mapping between the vectors $\dot{y}(0)$ and some vectors, which are called *tangent vectors to Q at q*.

Definition 3.6

1. The set of all tangent vectors to Q at point $q \in Q$ is called the tangent space $T_q Q$.
2. The union of all tangent spaces to Q is called the *tangent bundle* $TQ = \cup_{q \in Q} T_q Q$.

In other words, tangent space to a point $T_q Q$ defines the linear space of all tangent vectors to the given point q, sometimes called the "foot." Tangent spaces at different points q_1 and q_2 are *different* spaces as they are taken at different points—even though they may look similar. The easiest way to see that is to define the coordinates on the tangent bundle. A point on TQ can be written as (q, v), where $q \in Q$, and a vector $v \in \mathbb{R}^n$.

Thus, for example, TQ is not a vector space, simply because Q is, in general, not a vector space, so the expression $(q_1, v_1) + (q_2, v_2)$ does not make sense: $q_1 + q_2$ is not defined, for example, if Q is a sphere. A sketch of the tangent bundle to a sphere is shown in Fig. 3.1.

3.5 Action and Variations

Suppose a mechanical system has a configuration manifold Q. We consider a trajectory (a path) $\mathbf{q}(t)$, which is a smooth curve on Q for $t_0 < t < t_1$. We only need two derivatives of $\mathbf{q}(t)$ with respect to time to exist. Suppose this curve has to begin at $\mathbf{q}(t_0) = \mathbf{q}_0$ and end at $\mathbf{q}(t_1) = \mathbf{q}_1$. For every such curve $\mathbf{q}(t)$, we can also define $\dot{\mathbf{q}}(t)$ on the interval. If the Lagrangian $L(\mathbf{q}, \dot{\mathbf{q}}, t)$ is given as a function on the tangent bundle and time (i.e., L depends on $\mathbf{q}$, $\dot{\mathbf{q}}$ and t), we compute the following integral called *action*:

$$S = \int_{t_0}^{t_1} L(\mathbf{q}, \dot{\mathbf{q}}, t)dt . \tag{3.10}$$

Let us suppose that we deform the path $\mathbf{q}$, so it stays on Q, and it parameterized by an additional variable ϵ: $\mathbf{q}(t, \epsilon) \in Q$, $\mathbf{q}(t_0, \epsilon) = \mathbf{q}_0$, and $\mathbf{q}(t_1, \epsilon) = \mathbf{q}_1$. We assume

Fig. 3.1 A sketch representing the tangent bundle of a two-dimensional sphere S^2, which we denote TS^2

that at $\epsilon = 0$, the deformed path coincides with the original path, i.e., $\mathbf{q}(t, \epsilon = 0) = \mathbf{q}(t)$, and $\mathbf{q}(t, \epsilon)$ is differentiable with respect to ϵ.

Then, for a given ϵ, the function $\partial_\epsilon \mathbf{q}(t, \epsilon)$ is an element of the tangent space to $\mathbf{q}(t, \epsilon)$. The function $\partial_\epsilon \mathbf{q}(t, \epsilon)$ is so important that it merits its definition.

Definition 3.7 For any deformation $\mathbf{q}(t, \epsilon)$, the function (a curve in TQ)

$$\delta \mathbf{q} = \left. \frac{\partial \mathbf{q}(t, \epsilon)}{\partial \epsilon} \right|_{\epsilon=0} \tag{3.11}$$

is called the *variation* of the curve $\mathbf{q}(t) \in Q$.

The vertical line in Eq. (3.11) means that the derivative is taken first and then evaluated at $\epsilon = 0$. Clearly, there are many variations possible, depending on how the deformation of the curve is defined. We are going to derive principles of mechanics that are valid for *any* variations $\delta \mathbf{q}$.

3.6   Hamilton's Critical Action Principle

Let us now define the variation of the action given by Eq. (3.10). That variation is taken as the linear part in $\delta \mathbf{q}$ as follows:

$$\delta S = \left. \frac{\partial}{\partial \epsilon} \int_{t_0}^{t_1} L \left(\mathbf{q}(t, \epsilon), \dot{\mathbf{q}}(t, \epsilon), t \right) dt \right|_{\epsilon=0}. \tag{3.12}$$

Note that we take the derivative with respect to ϵ first, and *then* set $\epsilon = 0$, not the other way around! Then, we have the following:

Theorem 3.8 (Hamilton's Critical Action Principle) *The variation of action S defined by Eq. (3.12) has to vanish on the true trajectory* $\mathbf{q}(t)$ *for variations* $\delta\mathbf{q}$ *vanishing at the end points. In other words, the value of S is critical at the true trajectory* $\mathbf{q}(t)$ *for any* $\delta\mathbf{q}(t)$ *such that* $\delta\mathbf{q}(t_0) = \delta\mathbf{q}(t_1) = \mathbf{0}$. *In other words, on the true solution, we have*

$$\delta S = \delta \int_a^b L(\mathbf{q}, \dot{\mathbf{q}})dt = 0, \quad \delta\mathbf{q}(a) = \delta\mathbf{q}(b) = 0. \tag{3.13}$$

Let us compute the variation and see why Hamilton's principle makes sense. We take the derivatives with respect to ϵ in Eq. (3.12) as follows:

$$\delta S = \int_{t_0}^{t_1} \frac{\partial L}{\partial \mathbf{q}} \cdot \delta\mathbf{q}(t) + \underbrace{\frac{\partial L}{\partial \dot{\mathbf{q}}} \cdot \delta\dot{\mathbf{q}}(t)}_{\text{Integrate by parts}} \, dt \tag{3.14}$$

$$= \int_{t_0}^{t_1} \left(\boxed{\frac{\partial L}{\partial \mathbf{q}} - \frac{d}{dt}\frac{\partial L}{\partial \dot{\mathbf{q}}}} \right) \cdot \delta\mathbf{q}\, dt + \frac{\partial L}{\partial \dot{\mathbf{q}}} \cdot \delta\mathbf{q}\Big|_{t_0}^{t_1}.$$

The boundary terms cancel since $\delta\mathbf{q}(t_0) = \delta\mathbf{q}(t_1) = \mathbf{0}$. The boxed part is multiplied by an arbitrary function $\delta\mathbf{q}$. Here, we use a useful lemma.

Lemma 3.9 *Suppose* $\mathbf{f}(t)$ *is continuous, integrable on an interval* $[a, b]$. *If, for all integrable functions* $\mathbf{g}(t)$,

$$\int_a^b \mathbf{f}(t) \cdot \mathbf{g}(t)dt = 0, \quad then \quad \mathbf{f}(t) = 0. \tag{3.15}$$

We see that the boxed part in Eq. (3.14) must be equal to $\mathbf{0}$ since $\delta\mathbf{q}(t)$ is arbitrary. However, that boxed part is exactly (minus) Euler-Lagrange equations (Eq. (2.11)) with no external forces. Thus, Hamilton's principle leads the Euler-Lagrange equations.

One very important corollary of the critical action principle is as follows.

Lemma 3.10 (Lagrangian Connected by the Time Derivatives of Another Function) *Suppose two systems have Lagrangians* $L_1(\mathbf{q}, \dot{\mathbf{q}}, t)$ *and* $L_2(\mathbf{q}, \dot{\mathbf{q}}, t)$ *that are connected by a full-time derivative of some scalar function* $f(\mathbf{q}(t), t)$:

$$L_1(\mathbf{q}, \dot{\mathbf{q}}, t) = L_2(\mathbf{q}, \dot{\mathbf{q}}, t) + \frac{df(\mathbf{q}, t)}{dt}. \tag{3.16}$$

Then, these two systems have exactly the same Euler-Lagrange equations.

Proof Since $L_1(\mathbf{q}, \dot{\mathbf{q}}, t)$, and $L_2(\mathbf{q}, \dot{\mathbf{q}}, t)$ lead to the actions S_1 and S_2 that are related by

$$S_1 = S_2 + f(\mathbf{q}(b), t) - f(\mathbf{q}(a), t)$$

due to Eq. (3.10). The variation of the action is exactly the same since $\delta\mathbf{q}(a) = \delta\mathbf{q}(b) = 0$. Thus, these variations yield exactly the same Euler-Lagrange equations as Hamilton's principle applied to the original Lagrangian L_1.   ∎

Remark 3.11 (Minimizing the Action) Sometimes, Hamilton's principle is called "the principle of least action" or "minimal action principle." Sometimes the trajectory satisfying Eq. (3.13) is indeed minimizing the action, but finding a minimum of action S is not a requirement for a trajectory $\mathbf{q}(t)$ to be the true trajectory. Hamilton's principle (Eq. (3.13)) just states that S defined by Eq. (3.10) has a critical point at the true solution $\mathbf{q}(t)$, so variations of the action with respect to $\mathbf{q}$ are zero. No statement is made about S, as defined by Eq. (3.10), being minimized on the true solution. Proving that a trajectory $\mathbf{q}(t)$ actually minimizes the action is a much harder calculation and, most importantly, is not needed to find the true solution. We thus use the name of *critical action principle* for Eq. (3.13).

3.7 Worked Examples

To write equations of motion for a mechanical system, you need to do the following:

1. Identify the configuration manifold and introduce coordinates on the manifold and the tangent bundle, $(\mathbf{q}, \dot{\mathbf{q}})$.
2. It is advisable to compute the coordinates of individual "part" i of the problem, $\mathbf{x}_i(\mathbf{q}, t)$, and compute the kinetic energy of translation. For example, if $\mathbf{x}_i(\mathbf{q})$ is independent of time, we have

$$\dot{\mathbf{x}}_i = \frac{\partial \mathbf{x}_i}{\partial q^\alpha}\dot{q}^\alpha, \quad T = \frac{1}{2}\sum_{i,\alpha,\beta} m_i \frac{\partial \mathbf{x}_i}{\partial q^\alpha} \cdot \frac{\partial \mathbf{x}_i}{\partial q^\beta}\dot{q}^\alpha\dot{q}^\beta. \tag{3.17}$$

If appropriate, you will also need to compute the rotational kinetic energy. I strongly recommend performing this step; without this auxiliary step, it is easy to make a mistake.
3. Compute the kinetic and potential energies $T(\mathbf{q}, \dot{\mathbf{q}})$ and $V(\mathbf{q})$.
4. Define the Lagrangian $L = T - V$ and use the Euler-Lagrange equations (Eq. (2.11)).
5. Analyze the Euler-Lagrange equations as appropriate, for example, derive the conservation of energy, other integrals, phase portraits, and numerical solutions.

The examples below illustrate this procedure.

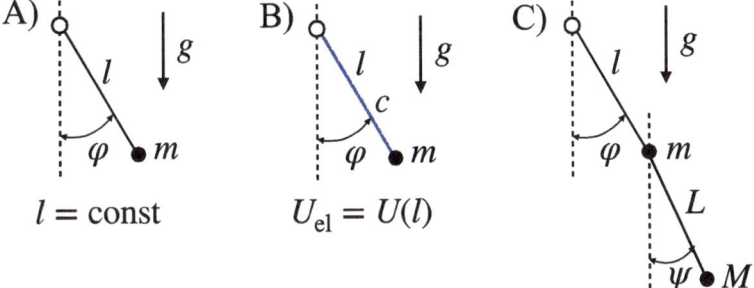

Fig. 3.2 Several examples related to the application of Lagrangian methods. Left (**a**): a classical pendulum. Middle (**b**): an elastic pendulum. The length of the pendulum l is a variable that contributes to the potential energy. Right (**c**): a double pendulum

Example 3.1 (Figure 3.2a. A Pendulum in 2D: A Mass m Attached with a Weightless Rod of Length l to a Given Point in Space)

1. Configuration manifold is a circle $Q = S^1$. The coordinate on the manifold is the angle φ, and the coordinates on the tangent bundle are $(\varphi, \dot{\varphi})$.
2. If φ is taken with respect to the vertical axis counterclockwise with $\varphi = 0$ pointing down, the coordinate of the bob is $\mathbf{x} = l(\sin\varphi, -\cos\varphi)$. The velocities are then given by $\dot{\mathbf{x}} = l\dot{\varphi}(\cos\varphi, \sin\varphi)$, and the kinetic energy is simply $T = \frac{1}{2}ml^2\dot{\varphi}^2$.
3. The potential energy is $V(\varphi) = -mgl\cos\varphi$, and the Lagrangian is

$$L = \frac{1}{2}ml^2\dot{\varphi}^2 + mgl\cos\varphi. \tag{3.18}$$

Euler-Lagrange equations are

$$ml^2\ddot{\varphi} + mgl\sin\varphi = 0, \quad \text{or} \quad \boxed{\ddot{\varphi} + \omega_0^2\sin\varphi = 0}, \quad \omega_0^2 := \frac{g}{l}. \tag{3.19}$$

The boxed equation is the familiar equation for a pendulum that you may have seen before.
4. The energy is given by

$$E = \dot{\varphi}\frac{\partial L}{\partial \dot{\varphi}} - L = T + V = \frac{1}{2}ml^2\dot{\varphi}^2 - mgl\cos\varphi, \tag{3.20}$$

and from general principles of energy conservation in Sect. 2.2, the energy E is conserved.

Example 3.2 (Figure 3.2b. Elastic Pendulum) Let us consider a slightly more complex case, where the pendulum's bob is held by an elastic string instead of a rigid rod. There is a potential energy $U(l)$ associated with the stretch and compression of the rod, for example, $U(l) = \frac{1}{2}K(l - l_0)^2$, where K is the string constant and l_0 is an equilibrium length. We will derive the equations of motion for arbitrary $U(l)$.

1. Configuration manifold is the product of the real line (for l) and a circle, $Q = \mathbb{R} \times S^1$, and $\mathbf{q} = (l, \varphi)$. We can also take $l \in [a, b]$ instead of the whole real line, which will not change the equations. The tangent bundle has coordinates $(l, \varphi, \dot{l}, \dot{\varphi})$.
2. The coordinate of the bob is given by $\mathbf{x} = l(t)(\sin\varphi(t), -\cos\varphi(t))$. Then,

$$\dot{\mathbf{x}} = \dot{l}(\sin\varphi, -\cos\varphi) + l\dot{\varphi}(\cos\varphi, \sin\varphi). \tag{3.21}$$

Notice that two vectors in parentheses in Eq. (3.21) have unit magnitude and are normal to each other. Then, $|\dot{\mathbf{x}}|^2 = \dot{l}^2 + l^2\dot{\varphi}^2$ and the kinetic energy is $T = \frac{1}{2}m(\dot{l}^2 + l^2\dot{\varphi}^2)$.
3. The Lagrangian is now

$$L = T - V = \frac{1}{2}m\left(\dot{l}^2 + l^2\dot{\varphi}^2\right) + mgl\cos\varphi - U(l). \tag{3.22}$$

Euler-Lagrange equations are

$$\begin{aligned}
\delta\varphi: \quad & m\frac{d}{dt}\left(l^2\dot{\varphi}\right) + mgl\sin\varphi = 0, \\
\delta l: \quad & m\ddot{l} - ml\dot{\varphi}^2 - mg\cos\varphi + U'(l) = 0.
\end{aligned} \tag{3.23}$$

4. The energy is conserved since the Lagrangian (Eq. (3.22)) is independent of time, i.e., $\frac{\partial L}{\partial t} = 0$:

$$E = \dot{l}\frac{\partial L}{\partial \dot{l}} + \dot{\varphi}\frac{\partial L}{\partial \dot{\varphi}} - L = \frac{1}{2}m\left(\dot{l}^2 + l^2\dot{\varphi}^2\right) - mgl\cos\varphi + U(l) = \text{const.} \tag{3.24}$$

Example 3.3 (Figure 3.2c. A Double Pendulum) A pendulum of the length l and bob's mass m has another pendulum of the length L and bob's mass M attached to it. This is a very famous example of a deceptively simple system that exhibits a surprisingly rich chaotic behavior [23].

1. Configuration manifold: $Q = S^1 \times S^1$.
2. The coordinates of masses m and M are $\mathbf{x}_m = l(\sin\varphi, -\cos\varphi)$, $\mathbf{x}_M = \mathbf{x}_m + L(\sin\psi, -\cos\psi)$.
3. The kinetic energy is

$$\begin{aligned}
T = &\frac{1}{2}m|\dot{\mathbf{x}}_m|^2 + \frac{1}{2}M|\dot{\mathbf{x}}_M|^2 = \frac{1}{2}ml^2\dot{\varphi}^2 \\
&+ \frac{M}{2}\left(l^2\dot{\varphi}^2 + L^2\dot{\psi}^2 + 2lL\dot{\varphi}\dot{\psi}\cos(\varphi - \psi)\right).
\end{aligned}$$

4. The potential energy is $U = -(m+M)gl \cos \varphi - MgL \cos \psi$, and the Lagrangian is

$$L = \frac{1}{2}(m + M)l^2\dot{\varphi}^2 + MLl\dot{\varphi}\dot{\psi}\cos(\varphi - \psi) + \frac{1}{2}ML^2\dot{\psi}^2$$
$$+ (m + M)gl \cos \varphi + MgL \cos \psi. \tag{3.25}$$

5. Since the Lagrangian is quadratic in velocities, the energy is

$$E = \frac{1}{2}(m + M)l^2\dot{\varphi}^2 + MLl\dot{\varphi}\dot{\psi}\cos(\varphi - \psi) + \frac{1}{2}ML^2\dot{\psi}^2$$
$$- (m + M)gl \cos \varphi - MgL \cos \psi. \tag{3.26}$$

One can see that in this case, the energy is conserved and is given as the kinetic plus potential energy.

6. Euler-Lagrange equations are then written as

$$\delta\varphi : \frac{d}{dt}\left[(m + M)l^2\dot{\varphi} + MLl\dot{\psi}\cos(\varphi - \psi)\right]$$
$$+ MLl\dot{\varphi}\dot{\psi}\sin(\varphi - \psi) + (m + M)gl \sin \varphi = 0$$
$$\delta\psi : \frac{d}{dt}\left[Ml^2\dot{\psi} + MLl\dot{\varphi}\cos(\varphi - \psi)\right] \tag{3.27}$$
$$+ MLl\dot{\varphi}\dot{\psi}\sin(\psi - \varphi) + MgL \sin \psi = 0.$$

The next two examples illustrate the difference between a system with a forced motion and a free motion of its parts.

Example 3.4 (Figure 3.3a. A Bead Freely Sliding on a Circular Hoop, with the Hoop Spinning with a Constant Angular Velocity ω) A bead on a hoop is spinning about the vertical axis with a *fixed* angular velocity ω. This problem is a bit tricky, as one is tempted to think that there are two angles: rotation of the hoop about its axis and inclination of the bead. However, the rotational angle of the hoop is *fixed*: if we call this angle θ, then $\theta = \omega t$. Thus, θ is not a variable; we need to know only φ at a given time to completely describe the system. We shall consider the alternative to this case below, where θ is a variable for the case when the hoop can rotate freely without friction.

1. The configuration manifold is S^1, with coordinate φ. The coordinates on the tangent bundle are $(\varphi, \dot{\varphi})$.
2. If the radius of the hoop is R, the coordinates of the bead in $\mathbb{R}^3$ and the corresponding velocities are

$$\mathbf{x} = R(\cos \omega t \sin \varphi, \sin \omega t \sin \varphi, -\cos \varphi),$$
$$\dot{\mathbf{x}} = R\omega(-\sin \omega t \sin \varphi, \cos \omega t \sin \varphi, 0) + R\dot{\varphi}(\cos \omega t \cos \varphi, \sin \omega t \cos \varphi, \sin \varphi). \tag{3.28}$$

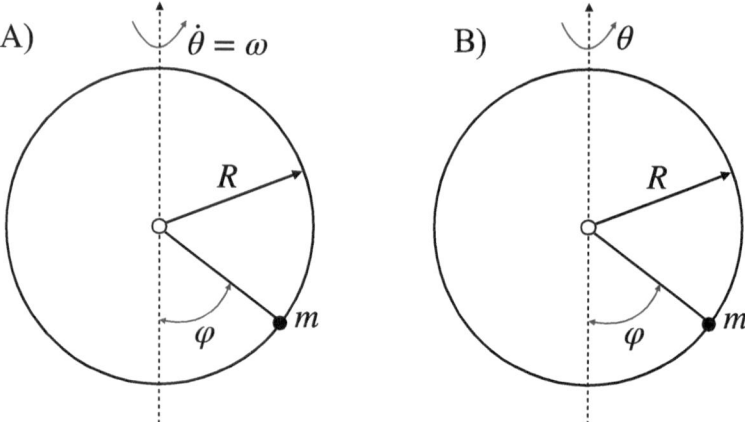

Fig. 3.3 Further examples related to the application of Lagrangian methods. Left (**a**): a bead on a hoop spinning with a fixed angular velocity ω. Right (**b**): the bead is sliding freely, and the hoop can also spin freely around the vertical axis going through its center

As we can see, this coordinate transformation is dependent on time. A somewhat lengthy calculation shows that $|\dot{\mathbf{x}}|^2 = R^2\omega^2 \sin^2\varphi + R^2\dot{\varphi}^2$. The kinetic energy is thus independent of time, $T = \frac{1}{2}m\left(R^2\omega^2\sin^2\varphi + R^2\dot{\varphi}^2\right)$. It is a bit of an accident since we have used the time-coordinate transformation $\mathbf{x} = \mathbf{x}(\mathbf{q}, t)$, and yet T came out to be independent of time, but there is a correction to the "physical energy" $T + V$ according to Eq. (2.17), due to the terms $|\partial_t\mathbf{x}(\mathbf{q}, t)|^2$.

3. The Lagrangian is $L = \frac{1}{2}mR^2\left(\omega^2\sin^2\varphi + \dot{\varphi}^2\right) + mgR\cos\varphi$ and the Euler-Lagrange equations are

$$mR^2\ddot{\varphi} + mgR\sin\varphi - mR^2\omega^2\sin\varphi\cos\varphi = 0, \quad \text{or}$$

$$\ddot{\varphi} + \sin\varphi(\omega_0^2 - \omega^2\cos\varphi) = 0, \quad \omega_0^2 := \frac{g}{R}. \tag{3.29}$$

4. The "energy" (or, more precisely, a constant of motion defined by Eq. (2.13)) is defined as

$$E = \dot{\varphi}\frac{\partial L}{\partial\dot{\varphi}} - L = \frac{1}{2}mR^2\left(\dot{\varphi}^2 - \omega^2\sin^2\varphi\right) - mgR\cos\varphi, \tag{3.30}$$

and, as one can see, E is not equal to $T + V$, since the transformation from $\mathbf{q}$ to $\mathbf{x}$ coordinates is time-dependent. Still, the energy (Eq. (3.30)) is conserved since the resulting Lagrangian is time-independent. Moreover, we observe from Eq. (3.29) that $\varphi = 0$ is always an equilibrium point. For $\omega > \omega_0$, there are two more equilibria at $\varphi_* = \pm\arccos\omega_0^2/\omega^2$. The equilibrium point at $\varphi = 0$ becomes unstable, and the point $\varphi = \varphi_*$ is stable. So, if the hoop is spun fast enough, the bead will move away from the bottom of the hoop to the other equilibrium.

Example 3.5 (Figure 3.3b. A Bead on a Circular Hoop, with the Hoop Spinning Freely About the Vertical Axis) The difference between this case and the case considered above is that the angle of rotation of the bead around the axis, θ, is no longer given by $\theta = \omega t$ but needs to have its own equation. At each point, the bead's coordinate is given by two angles: θ and φ. Then, we proceed as follows:

1. The configuration manifold is $T^2 = S^1 \times S^1$, with coordinates (φ, θ). The coordinates on the tangent bundle are $(\varphi, \theta, \dot{\varphi}, \dot{\theta})$.
2. The coordinates of the bead in $\mathbb{R}^3$ and the corresponding velocities are

$$\mathbf{x} = R(\cos\theta\sin\varphi, \sin\theta\sin\varphi, -\cos\varphi)$$

$$\dot{\mathbf{x}} = R\dot{\theta}(-\sin\theta\sin\varphi, \cos\theta\sin\varphi, 0) + R\dot{\varphi}(\cos\theta\cos\varphi, \sin\theta\cos\varphi, \sin\varphi)$$

$$(3.31)$$

The kinetic energy comes out surprisingly simple:

$$|\dot{\mathbf{x}}|^2 = R^2\sin^2\varphi\dot{\theta}^2 + R^2\dot{\varphi}^2,$$

$$T = \frac{1}{2}mR^2\left(\dot{\theta}^2\sin^2\varphi + \dot{\varphi}^2\right).$$

$$(3.32)$$

We are now ready to compute the equations of motion.

3. The Lagrangian is

$$L = T - V = \frac{1}{2}mR^2\left(\sin^2\varphi\dot{\theta}^2 + \dot{\varphi}^2\right) + mgR\cos\varphi.$$

$$(3.33)$$

We immediately notice that θ is a cyclic variable, so we expect a conservation law corresponding to $p_\theta = \frac{\partial L}{\partial \dot{\theta}} =$const. Indeed, let us divide the Lagrangian by mR^2, i.e., consider $\tilde{L} = L/(mR^2)$, and write the Euler-Lagrange equations for $\tilde{L}$, denoting again $\omega_0^2 = g/R$:

$$\delta\varphi : \ddot{\varphi} + \sin\varphi\left(\omega_0^2 - \dot{\theta}^2\cos\varphi\right) = 0,$$

$$\delta\theta : \frac{d}{dt}\left(\sin^2\varphi\dot{\theta}\right) = 0, \quad p_\theta = \dot{\theta}\sin^2\varphi = \text{const}.$$

$$(3.34)$$

We will study the deep generalizations of the cyclic variables and corresponding conservation laws, such as we got for p_θ later in the next chapter.

4. From general considerations, we know that since the transformation (Eq. (3.31)) does not depend on time explicitly, the energy (or, more precisely, energy-like quantity defined by Eq. (2.13)) is simply the mechanical energy and is conserved:

$$E = \dot{\theta}\frac{\partial L}{\partial \dot{\theta}} + \dot{\varphi}\frac{\partial L}{\partial \dot{\varphi}} - L = T + V$$

$$= \frac{1}{2}mR^2\left(\dot{\theta}^2\sin^2\varphi + \dot{\varphi}^2\right) - mgR\cos\varphi = \text{const}.$$

$$(3.35)$$

We can now substitute the conservation law for $\dot{\theta}$ from the second equation of Eq. (3.34) and obtain an effectively one-dimensional system for $\varphi(t)$:

$$E_1(\varphi, \dot{\varphi}) = \frac{1}{2}mR^2\left(\frac{p_\theta^2}{\sin^2\varphi} + \dot{\varphi}^2\right) - mgR\cos\varphi$$

$$= \frac{1}{2}mR^2\dot{\varphi}^2 + U_{\text{eff}}(\varphi) = \text{const}.$$

(3.36)

The solution curves on $(\varphi, \dot{\varphi})$ plane are level sets $E_1(\varphi, \dot{\varphi}) = $const. In principle, one could express $\dot{\varphi}$ as a function of φ from Eq. (3.36) and obtain the solution for $\varphi(t)$ implicitly in terms of quadratures, i.e., some integral of φ as a function of t:

$$\pm \int_0^\varphi \frac{\sqrt{mR^2}}{\sqrt{E_1 - U_{\text{eff}}(\varphi)}}d\varphi = t + C.$$

(3.37)

However, that solution is not very informative, so we will not study it further here.

3.8 Holonomic Constraints

Any constraint of the type $f(\mathbf{q}, t) = 0$ is called *holonomic*. Notice that if we differentiate this constraint, we formally get a constraint that involves time derivatives of $\mathbf{q}$. However, the important point in the definition of the holonomic constraint is that we *can* write the constraint as a function of coordinates. In this course, we will simply use the functions of the coordinates for holonomic constraints. In real life, you may work a little harder to understand that a constraint is holonomic.

Suppose a system has m constraints $f^\beta(\mathbf{q}, t) = 0$, $\beta = 1, \ldots, m$, with an n-dimensional configuration manifold Q. In reality, it means that only $n - m$ coordinates are independent, and the effective motion is on the $n - m$ dimensional manifold Q_C.

As we have seen from Theorem 3.4, a constraint, or a set of constraints in $\mathbb{R}^n$, defines a manifold. If there are nice and convenient coordinates on that configuration manifold, you should use them—it is almost always easier than considering the system with constraints (with a few exceptions). For example, for a pendulum, it is more convenient to use the angle, as we have done, rather than consider it as a problem with constraints.

However, finding convenient global coordinates on the manifold is not always feasible. The second method, using Lagrange multipliers, is less direct, but it will also give the reaction forces of the constraint. Let us show how it works by starting with Hamilton's principle with constraints:

$$S_\lambda = \int_{t_0}^{t_1} L(\mathbf{q}, \dot{\mathbf{q}}, t) + \lambda_\beta f^\beta(\mathbf{q}, t) dt, \quad \delta S_\lambda = 0, \quad f^\beta(\mathbf{q}, t) = 0, \beta = 1, \ldots, m$$

(3.38)

on all variations satisfying $\delta\mathbf{q}(t_0) = \delta\mathbf{q}(t_1) = \mathbf{0}$ We have introduced the vector of Lagrange multipliers $\boldsymbol{\lambda} = (\lambda_1, \ldots, \lambda_m)$. The variational principle, after one integration by parts of $\delta\dot{\mathbf{q}}$ term and cancelling the boundary terms, yields

$$\delta S_\lambda = \int_{t_0}^{t_1} \left(\boxed{\frac{\partial L}{\partial \mathbf{q}} - \frac{d}{dt}\frac{\partial L}{\partial \dot{\mathbf{q}}} + \lambda_\beta \frac{\partial f^\beta}{\partial \mathbf{q}}} \right) \cdot \delta\mathbf{q}\, dt$$

$$+ \int_{t_0}^{t_1} \delta\lambda_\beta f^\beta(\mathbf{q}, t) dt + \frac{\partial L}{\partial \dot{\mathbf{q}}} \cancel{\delta\mathbf{q}}\Big|_{t_0}^{t_1} = 0, \quad f^\beta(\mathbf{q}, t) = 0, \beta = 1, \ldots, m.$$

(3.39)

The vanishing of variations in $\delta\lambda_\beta$ simply enforces the constraints $f^\beta(\mathbf{q}, t) = 0$. The multiplier of $\delta\mathbf{q}$ (the boxed term) gives the modified Euler-Lagrange equations in the presence of constraints:

$$\frac{d}{dt}\frac{\partial L}{\partial \dot{\mathbf{q}}} - \frac{\partial L}{\partial \mathbf{q}} = \lambda_\beta \frac{\partial f^\beta}{\partial \mathbf{q}}.$$

(3.40)

The Lagrange multipliers λ_β need to be computed from the fact that the constraints are satisfied. The right-hand side of Eq. (3.40) yields the reaction force of the constraint, forcing the system to obey the constraint at all times. Since $\frac{\partial f^\beta}{\partial \mathbf{q}}$ is normal to the surface $f^\beta = 0$, the reaction force is also normal to that surface. Since the mechanical system is moving on the intersection of surfaces $f^\beta = 0$, *the constraint forces perform no work.*

Rolling Without Slipping In general, rolling without slipping in 3D introduces a non-holonomic constraint in the system, and we shall postpone the discussion of such systems until Chap. 11. However, rolling without slipping in 2D gives a holonomic constraint: it connects the path traveled along the surface with the angle of revolution. For the motion of a rigid cylinder of radius R along a curve, the path by the contact point is simply $R\phi$. If the disk is moving in contact with a straight line (e.g., on a table), that is also the distance traveled by the center of the disk. If the substrate is curved, e.g., is a circle, then the motion of the center of the disk has to be computed from elementary geometry.

In solving practical problems, it is advantageous to describe the motion of the rolling disk by one variable, for example, rotation angle φ, and compute the motion of the center of mass $\mathbf{x}(\varphi)$. The kinetic energy then consists of the translational and rotational parts:

$$T = \frac{1}{2}m|\dot{\mathbf{x}}|^2 + \frac{1}{2}I\dot{\varphi}^2 = \frac{1}{2}\left(m\left|\frac{\partial\mathbf{x}}{\partial\varphi}\right|^2 + I\right)\dot{\varphi}^2, \tag{3.41}$$

where I is the moment of inertia computed about the center of mass.

3.9 Lagrange-d'Alembert's Principle for External Forces and Rayleigh's Dissipation Function

Suppose now that there are external or nonconservative forces acting on the system with Lagrangian L. The variational principle should result in the Euler-Lagrange equations (Eq. (2.11)) incorporating external nonconservative forces $\mathbf{Q} = (Q_1, \ldots Q_n)$. Lagrange-d'Alembert's variational principle states that

$$\delta S = \delta \int_{t_0}^{t_1} L(\mathbf{q}, \dot{\mathbf{q}}, t)\mathrm{d}t = -\int \mathbf{Q} \cdot \delta \mathbf{q} \mathrm{d}t. \tag{3.42}$$

Thus, the variation of δS is not zero but is dependent on the external force. The expression on the right-hand side is called the *virtual work*. One way to compute that force $\mathbf{Q}$ is directly from (2.9) using the force acting on each particle $\mathbf{Q}_i(\mathbf{x}_i, \dot{\boldsymbol{\xi}}, t)$.[5] Another way to derive that force $\mathbf{Q}$ is to use the *generalized force equality*, leading to the same result:

$$\mathbf{Q} \cdot \delta \mathbf{q} = \sum_l \mathbf{Q}_i(\mathbf{x}_i, \dot{\mathbf{x}}_i, t) \cdot \delta \mathbf{x}^i = \underbrace{\sum_i \mathbf{Q}_i(\mathbf{x}_i(\mathbf{q}), \dot{\mathbf{x}}_i(\mathbf{q}, \dot{\mathbf{q}}), t) \cdot \frac{\partial \mathbf{x}_i}{\partial \mathbf{q}}}_{=\mathbf{Q}(\mathbf{q}, \dot{\mathbf{q}}, t)} \cdot \delta \mathbf{q}. \tag{3.43}$$

We thus have the dissipative Euler-Lagrange equations written as

$$\frac{d}{dt}\frac{\partial L}{\partial \dot{q}^\alpha} - \frac{\partial L}{\partial q^\alpha} = Q_\alpha = \sum_i \mathbf{Q}_i(\mathbf{x}_i(\mathbf{q}), \dot{\mathbf{x}}_i(\mathbf{q}, \dot{\mathbf{q}}), t) \cdot \frac{\partial \mathbf{x}_i}{\partial q^\alpha}. \tag{3.44}$$

Equation (3.43) is, in principle, enough to compute the dissipative forces; however, that computation may be quite complicated. Our goal is to show that *when the friction force $\mathbf{Q}_i$ is linear in velocities*, every component of the generalized friction

---

[5] We denote $\mathbf{Q}(\mathbf{q}, \dot{\mathbf{q}}, t)$ to be generalized force with the component α corresponding to Q_α on the right-hand side of the Euler-Lagrange equations. In contrast, $\mathbf{Q}_i$ is the nonconservative force acting on the particle i.

force Q_α can be derived as a gradient of some scalar function with respect to $\dot{q}^\alpha$. We will call that function the *Rayleigh's dissipation function*.

To proceed, let us now compute the generalized forces in a particular case when the force is called by friction with outside media. In that case, $\mathbf{Q}_i = -\mathbb{A}_i \dot{\mathbf{x}}_i = -\mathbb{A}_i \mathbf{v}_i$ (no sum), where $\mathbb{A}_i$ may depend on $\mathbf{q}$ or t, but not on $\dot{\mathbf{q}}$. We again use the *cancellation of dots*; see also Eq. (2.5):

$$\mathbf{v}_i = \dot{\mathbf{x}}_i = \frac{\partial \mathbf{x}_i}{\partial q^\alpha}\dot{q}^\alpha + \frac{\partial \mathbf{x}_i}{\partial t} \quad \Rightarrow \quad \frac{\partial \mathbf{v}_i}{\partial \dot{q}^\alpha} = \frac{\partial \dot{\mathbf{x}}_i}{\partial \dot{q}^\alpha} = \frac{\partial \mathbf{x}_i}{\partial q^\alpha}. \tag{3.45}$$

Then, due to Eqs. (3.43) and (3.45), we can write

$$Q_\alpha = -\sum_i \mathbb{A}_i \mathbf{v}_i \cdot \frac{\partial \mathbf{v}_i}{\partial \dot{q}^\alpha}. \tag{3.46}$$

Sometimes, $\mathbb{A}_i$ has a particularly simple form, and Eq. (3.46) can be written as a gradient of a certain function with respect to $\dot{q}^\alpha$. For that to happen, $\mathbb{A}$ must be symmetric and cannot depend on $\dot{\mathbf{q}}$. However, $\mathbb{A}$ can depend on time and the coordinates $\mathbf{q}$. One particular example is when $\mathbb{A}_i = a_i \mathbb{I}$ (the identity matrix), with a_i being constants:

$$\mathcal{R} = \frac{1}{2}\sum_i a_i |\mathbf{v}^i|^2 = \frac{1}{2}\sum_i a_i \left| \frac{\partial \mathbf{x}_i}{\partial \mathbf{q}}\dot{\mathbf{q}} + \frac{\partial \mathbf{x}_i}{\partial t} \right|^2, \quad Q_\alpha = -\frac{\partial \mathcal{R}}{\partial \dot{q}^\alpha}. \tag{3.47}$$

We chose the sign, so $\mathcal{R}$ in Eq. (3.47) is positive definite. The function $\mathcal{R}$ is called Rayleigh's dissipation function. Notice that more complex cases can be treated by this method as well, for example, when a_i are functions of velocities only, so Rayleigh's function approach is more powerful than the linear friction case (Eq. (3.47)). Of course, in that case, the expressions for Rayleigh's function $\mathcal{R}$ also become more complex.

A Damped Pendulum As a simple example, we consider the motion of a 2D pendulum with the additional friction of the bob with air given by $-\beta\dot{\mathbf{x}}$. All the other parameters remain as above. Using the notation above, the speed of the bob is given by $\dot{\mathbf{x}} = l\dot{\varphi}(\cos\varphi, \sin\varphi)$. Rayleigh's function is $\mathcal{R} = \frac{1}{2}\beta l^2 \dot{\varphi}^2$, and the Euler-Lagrange equations for pendulum (Eq. (3.19)) for the case of friction are

$$\ddot{\varphi} + \omega_0^2 \sin\varphi = -\kappa\dot{\varphi}, \quad \kappa = \frac{\beta}{m}. \tag{3.48}$$

The energy is no longer conserved but dissipated:

$$\dot{E} = -\beta l^2 \dot{\varphi}^2 = -2\mathcal{R} \le 0. \tag{3.49}$$

Note that the rate of dissipation of energy is proportional to Rayleigh's function, as expected.

3.10 Looking Forward

We have just touched upon a more rigorous definition of analytical mechanics that forms the very foundation of modern mechanics. It is perhaps not an exaggeration to say that the progress in mechanics has been based on the successful application of modern mathematical methods, bringing clarity to otherwise complex problems. There are many excellent books that study the geometric background underpinning modern mechanics, such as books by José and Saletan [5], Holm [2, 3], Marden and Ratiu [4], Arnol'd [1], and many others. You will find a lot of interesting applications and modern results in classical mechanics in [24]. We will keep providing deeper insights into the mathematical theory without going into too much abstraction.

Regarding Rayleigh's dissipation function, there is an interesting discussion in [25] about the use of that theory for friction forcing that is nonlinear in velocities. A more general and comprehensive exposition of Rayleigh's dissipation function for general forces is considered in [26], which no doubt will interest a reader curious about treating problems with friction from the geometric points of view.

3.11 Practice Problems

1. A vertical wheel rolls freely without slipping on a horizontal table. The moment of inertia of the wheel is I, mass is M, and radius is R. In the wheel, there is a hamster running with a constant speed V. The hamster has the mass m, and it is separated from the wheel's edge by height h, so the hamster's center of mass is always $R - h$ away from the wheel's center. See Fig. 3.4, left panel.

   Find the configuration manifold of the problem and write the Euler-Lagrange equations. Find the energy E. Is the energy conserved, and is it equal to the kinetic+potential energy?

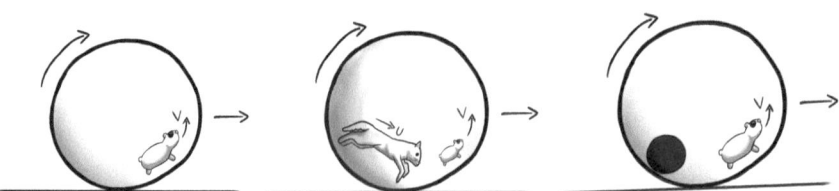

Fig. 3.4 Left: a hamster in a rolling wheel. Center: a hamster chased by a cat. Right: a hamster and a ball in the wheel

Are there any equilibria for some values of parameters? If yes, find them and
describe the motion about the equilibria (stable or unstable). You don't need to do
stability computations here. *Hint* Use the energy method to describe the motion
about the equilibria.

2. The same as in Problem 1, only the hamster is chased by a cat who is running on
   a parallel track to the hamster, as shown in Fig. 3.4, center panel. The masses of
   the hamster and cat are m_1 and m_2, the distances from the edge of the wheel to
   their centers of mass are h_1 and h_2, and the velocities are v_1 and v_2, respectively.
   Since the tracks are parallel, when the hamster and the cat meet, they just pass
   through with no interaction.
   Are there any equilibria (steady states) for some values of parameters? If yes,
   find them and describe the motion about the equilibria (stable or unstable). You
   don't need to do stability computations here. Find the energy E. Is the energy
   conserved, and is it equal to the kinetic+potential energy?

3. The same as in Problem 2, only instead of a cat, there is an armadillo rolled in a
   ball, rolling freely without slipping inside the wheel, as shown in Fig. 3.4, right
   panel. The masses of the hamster and armadillo are m_1 and m_2; the distances
   from the edge of the wheel to their centers of mass are h_1 and h_2, respectively;
   and the velocity of the hamster is V (fixed). Since the tracks are parallel, when
   the hamster and armadillo meet, they just pass through with no interaction.
   Find the energy E. Is the energy conserved, and is it equal to the
   kinetic+potential energy? Are there any equilibria for some values of
   parameters? If yes, find them and describe the motion about the equilibria
   (stable or unstable). You don't need to do stability computations here. In this
   problem, you only are only asked to consider the equilibrium where both the
   hamster and armadillo are down. (You will be able to compute more general
   stability criteria later in the course in Sect. 5.2).

4. **Euler-Lagrange equations for dissipative systems?** Consider the equation for
   a damped oscillator in one dimension:

$$\ddot{q} + \gamma \dot{q} + U'(q) = 0, \tag{3.50}$$

where $\gamma > 0$ is a constant, and $U(q)$ is the potential energy, assumed to be a
convex function, so the solution is stable.

(a) Consider the "traditional energy" $E_0 = \frac{1}{2}\dot{q}^2 + U(q)$ (kinetic plus potential
    energy), and compute $\frac{dE_0}{dt}$ when q is a solution of Eq. (3.50). Show that
    $\frac{dE_0}{dt} \leq 0$ on all solutions, so the system is dissipative.

(b) Show that Eq. (3.50) can be multiplied by a certain function $\mu(t)$ such that it
    becomes the Euler-Lagrange equation for some time-dependent Lagrangian.
    Hint Exponentials are a good choice in this case.

(c) Compute the "energy in the sense of this course" $E = \dot{q}\frac{\partial L}{\partial \dot{q}} - L$ and show
    how E is related to E_0.

(d) Compute $\frac{dE}{dt}$ and show it is consistent with the general expression $\dot{E} = -\frac{\partial L}{\partial t}$.

5. Compute Rayleigh's function for 2D elastic pendulum, assuming that the force of friction on the bob is equal to $F_\mathbf{x} = -\alpha\dot{\mathbf{x}}$, where $\alpha > 0$ is a constant and $\mathbf{x}$ is the coordinate of the bead in $\mathbb{R}^2$. Compute Rayleigh's function and the generalized forces of friction in the Euler-Lagrange equations (Eq. (3.23)). Compute the rate of change of energy. Is the system dissipative, i.e., is $\dot{E} \leq 0$?

6. For the bead on the hoop spinning with a constant angular velocity considered above, consider the friction of the bead with the outside air given by $F_\mathbf{x} = -\alpha\dot{\mathbf{x}}$, where $\alpha > 0$ is a constant and $\mathbf{x}$ is the coordinate of the bead in $\mathbb{R}^3$. Compute Rayleigh's function and the generalized force of friction in the Euler-Lagrange equations (Eq. (3.29)). Is the system dissipative, i.e., is $\dot{E} \leq 0$?

Chapter 4
Noether's Theorem and Conservation Laws

4.1 Noether's Theorem

Cyclic Coordinates If a Lagrangian does not depend on the coordinate q^α for some α (but, of course, depends on $\dot{q}^\alpha$—otherwise, that coordinate is irrelevant), this coordinate is called *cyclic*. The important property of a cyclic coordinate is that the corresponding momentum $\frac{\partial L}{\partial \dot{q}^\alpha}$ is conserved if there are no external forces. Indeed, from Euler-Lagrange equations, for the coordinate α *only*, we get

$$\frac{d}{dt}\frac{\partial L}{\partial \dot{q}^\alpha} - \frac{\partial L}{\partial q^\alpha} = 0, \quad \Rightarrow \quad \frac{\partial L}{\partial \dot{q}^\alpha} = \text{const} \tag{4.1}$$

You have already observed the cyclic coordinate in the Lagrangian for Kepler's problem (Eq. (2.20)), with the cyclic coordinate being θ, leading to the conservation of momentum $mr^2\dot{\theta}$ as given in Eq. (2.24).

Noether's theorem is a way to compute the constants of motion similar to that of the momentum to cyclic coordinate whenever there is a continuous symmetry of the Lagrangian. Noether's theorem is one of the most fundamental and beautiful concepts in mechanics, stating that for every continuous symmetry of the system, there is a conserved quantity. The reverse is, in general, not true: some conserved quantities of mechanical systems may not arise from any symmetry (or that symmetry is very hard to find).

Suppose there is a one-parameter transformation of coordinates $\mathbf{Q} = \mathbf{Q}(\mathbf{q}, t, \epsilon)$, defining the time derivative $\dot{\mathbf{Q}} = \partial_t \mathbf{Q}(\mathbf{q}, t, \epsilon)$, with ϵ being a parameter. Suppose that at $\epsilon = 0$ we have $\mathbf{Q} = \mathbf{q}$, and the transformations are invertible, i.e., we can write $\mathbf{q} = \mathbf{q}(\mathbf{Q}, t, \epsilon)$ and $\dot{\mathbf{q}} = \dot{\mathbf{q}}(\mathbf{Q}, \dot{\mathbf{Q}}, t, \epsilon)$.

© The Author(s), under exclusive license to Springer Nature Switzerland AG 2025
V. Putkaradze, *A Concise Introduction to Classical Mechanics*,
Surveys and Tutorials in the Applied Mathematical Sciences 16,
https://doi.org/10.1007/978-3-031-84977-0_4

Define the new Lagrangian

$$L_\epsilon = L(\mathbf{q}(\mathbf{Q}, t, \epsilon), \dot{\mathbf{q}}(\mathbf{Q}, \dot{\mathbf{Q}}, t, \epsilon), t). \tag{4.2}$$

Notice that, unlike our derivation of the Euler-Lagrange equations, the deformation $\mathbf{q}(\mathbf{Q}, t, \epsilon)$ is caused by the symmetry, not arbitrary deformation of the path along the configuration manifold Q. Thus, it is a particular case of choosing $\delta\mathbf{q}$. The calculation is formally the same, but the meaning is different. We are interested in the case when L_ϵ is invariant with respect to the transformation, which means that L_ϵ does not depend on ϵ.

Consider $S_\epsilon = \int_{t_0}^{t_1} L_\epsilon dt$. Define, as before, δS to be the derivative of S_ϵ evaluated at $\epsilon = 0$. Since we assume that L_ϵ does not depend on ϵ, $\delta S = 0$. That is the consequence of the invariance, not Hamilton's critical action principle. Then, we have

$$\delta S = \int_{t_0}^{t_1} \left(\frac{\partial L}{\partial \mathbf{q}} \cdot \delta\mathbf{q}(t) + \frac{\partial L}{\partial \dot{\mathbf{q}}} \cdot \delta\dot{\mathbf{q}} \right) dt = \text{(Integrate last term by parts)}$$

$$= \int_{t_0}^{t_1} \underbrace{\left(\frac{\partial L}{\partial \mathbf{q}} - \frac{d}{dt} \frac{\partial L}{\partial \dot{\mathbf{q}}} \right)}_{= 0 \text{ by E-L eqs}} \cdot \delta\mathbf{q}\, dt + \boxed{\left. \frac{\partial L}{\partial \dot{\mathbf{q}}} \cdot \delta\mathbf{q} \right|_{t_0}^{t_1}} = 0 \tag{4.3}$$

Now, we use the fact that Euler-Lagrange equations cancel the term multiplying $\delta\mathbf{q}$. Thus, the boundary (boxed) terms must cancel, so the quantity

$$N = \frac{\partial L}{\partial \dot{\mathbf{q}}} \cdot \delta\mathbf{q} \tag{4.4}$$

must be the same for the initial and end points. Since these points are arbitrary, the quantity N defined by Eq. (4.4) must be conserved.

The above result (Eq. (4.4)) relates the symmetry of the coordinates to the conservation law. There is an additional statement for the case when the transformation involves time. Suppose the transformation given by $t = t(\mathbf{Q}, \tau, \epsilon)$, $\mathbf{q} = \mathbf{q}(\mathbf{Q}, t, \epsilon)$ leaves the Lagrangian invariant. We form the new Lagrangian in the extended configuration manifold $Q \times \mathbb{R}$, with the additional coordinate in $\mathbb{R}$ being the time variable. The velocities transform as $\dot{q} \rightarrow \frac{d\mathbf{Q}/d\tau}{dt/d\tau}$. Then, we define the new Lagrangian as

$$\tilde{L}_\epsilon = L\left(\mathbf{Q}, \frac{d\mathbf{Q}/d\tau}{dt/d\tau}, t \right) \frac{dt}{d\tau} \tag{4.5}$$

We then consider the infinitesimal extended symmetry to include time $(\delta\mathbf{q}, \delta t)$ and apply this result to the new Lagrangian $\tilde{L}_\epsilon$ given by Eq. (4.5). We will need to additionally differentiate the Lagrangian with respect to $\frac{dt}{d\tau}$ and then set $\epsilon = 0$, so $\frac{dt}{d\tau} = 1$ *after* the differentiation. After some calculation, tracing the boundary terms as in Eq. (4.3) gives the conserved quantity:

$$N = \frac{\partial L}{\partial \dot{\mathbf{q}}} \cdot \delta \mathbf{q} - \left(\dot{\mathbf{q}} \cdot \frac{\partial L}{\partial \dot{\mathbf{q}}} - L \right) \delta t = \frac{\partial L}{\partial \dot{\mathbf{q}}} \cdot \delta \mathbf{q} - Et . \qquad (4.6)$$

Note that the term in parentheses multiplying δt in (4.6) is the energy-like quantity E we defined earlier in Eq. (2.13).

That quantities N in Eqs. (4.4) and (4.6) are called *Noether's integrals* and are due to Emmy Noether. She discovered these quantities in her groundbreaking work in 1918; see modern translation of her work [27]. Her result is fundamental in modern physics and mathematics.

4.2 Examples

Cyclic Coordinates and Conservation of Linear Momenta If the Lagrangian does not depend on q^α, the one-parameter transformation is $q^\alpha = Q^\alpha + \epsilon$, with all other q's remaining constant. Then $\delta \mathbf{q} = \mathbf{e}_\alpha$, the basis vector number α. Noether's integral is

$$N = p_\alpha = \frac{\partial L}{\partial \dot{q}^\alpha} . \qquad (4.7)$$

For an isolated system, the potential energy does not depend on any coordinates $\mathbf{x}^i$, so all the components of linear momentum are conserved.

Angular Momentum Consider the system moving in $\mathbb{R}^3$, with $\mathbf{q} = \mathbf{x} \in \mathbb{R}^3$. Suppose the Lagrangian is invariant with respect to rotation about a given axis $\mathbf{n}$. Then, if $R(\epsilon, \mathbf{n})$ is the rotation matrix about this axis by the angle ϵ, then $\mathbf{x} = R(\epsilon, \mathbf{n})\mathbf{X}$, and $\delta \mathbf{x} = \mathbf{n} \times \mathbf{x}$. Noether's integral is

$$N = \frac{\partial L}{\partial \dot{\mathbf{x}}} \cdot \mathbf{n} \times \mathbf{x} = \mathbf{n} \cdot \left(\mathbf{x} \times \frac{\partial L}{\partial \dot{\mathbf{x}}} \right) = \mathbf{n} \cdot (\mathbf{x} \times \mathbf{p}) = \mathbf{n} \cdot \mathbf{L} . \qquad (4.8)$$

Thus, the projection of the angular momentum $\mathbf{L}$ on the axis of symmetry $\mathbf{n}$ of the system is conserved. For an isolated system, all coordinates of angular momentum $\mathbf{L}$ are conserved.

Time Shift For a system independent of time, the symmetry is $t' = t + \epsilon$ and $\delta t = 1$, $\delta \mathbf{q} = \mathbf{0}$. Noether's integral is then simply E, the energy-like integral we derived earlier in Eq. (2.13).

A More General One-Parameter Transformation Consider the Lagrangian

$$L = \frac{1}{2}|\dot{\mathbf{q}}|^2|\mathbf{q}|^4. \tag{4.9}$$

Here, $\varphi(\xi)$ is a given smooth function of the scalar argument ξ. For simplicity, we assume $\varphi(\xi)$ to be monotonic in ξ.

1. There is an obvious time shift symmetry $t \rightarrow t + \epsilon$, so the energy is conserved:

$$E = \dot{\mathbf{q}} \cdot \frac{\partial L}{\partial \dot{\mathbf{q}}} - L = \frac{1}{2}|\dot{\mathbf{q}}|^2|\mathbf{q}|^4 = L = \text{const}. \tag{4.10}$$

2. If $\mathbf{q} \in \mathbb{R}^3$, there is also rotational symmetry, so the angular momentum

$$\mathbf{l} = \mathbf{q} \times \frac{\partial L}{\partial \dot{\mathbf{q}}} = (\mathbf{q} \times \dot{\mathbf{q}})|\mathbf{q}|^4 \tag{4.11}$$

   is also conserved.
3. There is, however, another symmetry that combines the dilation in coordinates and time. We look for transformations in the following form:

$$\mathbf{Q} = e^\epsilon \mathbf{q}, \quad \tau = e^{a\epsilon}t. \tag{4.12}$$

We need to find a such that the quantity (integrand of the action) $L(\mathbf{q}, \dot{\mathbf{q}}, t)\mathrm{d}t$ remains invariant with respect to the transformation (Eq. (4.12)). Note that we have to multiply the Lagrangian by $\frac{\mathrm{d}t}{\mathrm{d}\tau}$ according to Eq. (4.5); otherwise, we will obtain an incorrect result. That quantity $L\mathrm{d}t$ is invariant when

$$L(\mathbf{Q}, \dot{\mathbf{Q}}, \tau)\frac{\mathrm{d}t}{\mathrm{d}\tau} \rightarrow e^{(6-a)\epsilon}L(\mathbf{q}, \dot{\mathbf{q}}, t)$$

giving $a = 6$. Then, from Eq. (4.12), we obtain $\delta\mathbf{q} = \mathbf{q}$, $\delta t = 6t$, and Noether's integral is

$$N = \delta\mathbf{q} \cdot \frac{\partial L}{\partial \dot{\mathbf{q}}} - E\,\delta t = (\mathbf{q} \cdot \dot{\mathbf{q}})|\mathbf{q}|^4 - 6E\,t. \tag{4.13}$$

Noether's integral (Eq. (4.13)) looks strange, having an explicit dependence on time. However, one can readily verify that $N = \text{const}$ and we did not make a mistake. This verification is done by taking the time derivative of Eq. (4.13), using the fact that E is constant, and utilizing the Euler-Lagrange equations for the Lagrangian (Eq. (4.9)).

4.3 Invariance of Dissipative Systems

Suppose the system is dissipative and has Rayleigh's dissipation function, and assume that both Lagrangian and Rayleigh's functions are invariant with respect to some symmetry $\mathbf{q} = \mathbf{q}(\mathbf{Q}, t, \epsilon)$. Is Noether's integral (4.4) still conserved? The answer is no, as the following example shows.

Take the Lagrangian L and Rayleigh's function $\mathcal{R}$ for a system in $\mathbf{x} = (x^1, x^2) \in \mathbb{R}^2$ to be of the following form:

$$L = \frac{1}{2}|\dot{\mathbf{x}}|^2 - V(r), \quad r = |\mathbf{x}|, \quad \mathcal{R} = \frac{1}{2}K|\dot{\mathbf{x}}|^2. \tag{4.14}$$

That is a particle moving in a central potential with symmetric dissipation. Lagrangian and Rayleigh's functions are invariant with respect to rotation:

$$\begin{pmatrix} x^1 \\ x^2 \end{pmatrix} = \begin{pmatrix} \cos\epsilon & \sin\epsilon \\ -\sin\epsilon & \cos\epsilon \end{pmatrix} \begin{pmatrix} X^1 \\ X^2 \end{pmatrix}, \quad \delta\mathbf{x} = \begin{pmatrix} x^2 \\ -x^1 \end{pmatrix}. \tag{4.15}$$

Noether's integral is then $\Gamma = x^2\dot{x}^1 - x^1\dot{x}^2$. The equations of motion are

$$\ddot{\mathbf{x}} - \frac{V'(r)}{r}\mathbf{x} = -K\dot{\mathbf{x}} \tag{4.16}$$

The evolution equation for Γ is

$$\dot{\Gamma} = x^2\ddot{x}^1 - x^1\ddot{x}^2 = -Kx^2\dot{x}^1 + Kx^1\dot{x}^2 = -K\Gamma, \quad \Rightarrow \quad \Gamma = \Gamma_0 e^{-Kt}. \tag{4.17}$$

(Note that x^1 and x^2 in Eq. (4.17) are indices of the components, not powers of x). Thus, Γ is not conserved as one would expect by naively applying Noether's theorem to dissipative systems.

4.4 Looking Forward

Since its discovery, Noether's theorem has played a fundamental role in mechanics. That wonderful theorem offers a unique way to compute extra the constants of motion as long as at least one continuous symmetry is known. One of the most comprehensive, authoritative, and thorough treatments of symmetries applied to general systems of differential equations is [28]. The description of Noether's theorem is contained in Sect. 4.4, but a reader will find a wealth of information related to applications of symmetries to variational and non-variational problems in that classical text.

There have been numerous studies concerning the extension of Noether's theorem for nonconservative systems. A reader interested in learning more about recent developments should read [29–31]. For recent developments related to applications of Noether's theorem to dissipative problems in quantum mechanics, we refer the reader to [32, 33].

4.5   Practice Problems

1. Suppose a certain system with two degrees of freedom has the Lagrangian

$$L = \dot{q}^1 \dot{q}^2 - \frac{1}{2}(q^1 q^2)^2.$$

   (a) Compute the Euler-Lagrange equations for the system.
   (b) Show that the Lagrangian is invariant with respect to the symmetry $q^1 \to e^{\epsilon} q^1$, $q^2 \to e^{-\epsilon} q^2$, and compute Noether's integral:

$$\Gamma = \frac{\partial L}{\partial \dot{q}} \cdot \delta \mathbf{q} = \dot{q}^2 q^1 - \dot{q}^1 q^2, \quad \delta \mathbf{q} := \left. \frac{\partial \mathbf{q}}{\partial \epsilon} \right|_{\epsilon=0}. \tag{4.18}$$

2. Suppose the Lagrangian

$$L(\mathbf{q}, \dot{\mathbf{q}}) = \frac{1}{2} \left(\dot{q}_1^2 + \dot{q}_2^2 \right) - U(q_1, q_2) \quad \mathbf{q} \in \mathbb{R}^2$$

   has the first integral

$$\Gamma = q_1 \dot{q}_2 - q_2 \dot{q}_1.$$

   (We write indices down in this particular case so as not to confuse them with powers of q_i). Find the general form of the function $U(q_1, q_2)$ for which such an integral Γ is possible. Show that this integral can be obtained by applying Noether's theorem.

3. Suppose $\mathbf{q} \in \mathbb{R}^3$ and consider the (unphysical) Lagrangian

$$L = \frac{1}{2|\mathbf{q}|^2} |\dot{\mathbf{q}}|^2. \tag{4.19}$$

   (a) Find the Euler-Lagrange equations for the Lagrangian (4.19).
   (b) Show that the Lagrangian is invariant with respect to dilations:

$$q_i \to e^{\epsilon} \mathbf{q}, \quad i = 1, 2, 3.$$

   Find Noether's integral Γ.

(c) Alternatively, you can show that this Lagrangian is invariant with respect to rotations:

$$\mathbf{q} \to A(\epsilon, \mathbf{n})\mathbf{q},$$

where $A(\epsilon, \mathbf{n})$ is the rotation matrix about a given axis $\mathbf{n}$ by angle ϵ. For example, take $\mathbf{n} = \mathbf{e}_3$, the unit vector in $\mathbf{z}$ direction, and compute the transformation explicitly. Then, find Noether's integral and show that it is equal to the angular momentum in the $\mathbf{e}_3$ direction.

Chapter 5
Linear Stability of Small Oscillations About an Equilibrium

5.1 A Review in Linear Algebra

Matrices An $n \times m$ *matrix* is a $n \times m$ table of numbers. For a matrix $\mathbb{A}$, the selection of (i, j)-th element produces a number a_{ij}, so $\mathbb{A} = \left\{ a^i_j \right\}_{i=1,\ldots,n;\, j=1,\ldots m}$. We will denote the process of element selection as $(\mathbb{A})^i_j = a^i_j$. The elements a^i_j are called entries of the matrix. A matrix is called symmetric if the transpose of the matrix equals to itself: $\mathbb{A}^T = \mathbb{A}$, or $a^i_j = a^j_i$ (we are a bit loose here with swapping the coordinates).

Multiplication of Matrices and Vectors Vectors are $m \times 1$ matrices (columns). An $n \times m$ matrix can be viewed as n vectors of length m put together, i.e., a row of $m \times 1$ vectors. Alternatively, it can be viewed as a vector of rows of length n.

Let us first derive the multiplication of rows and vectors. For a row of length m, $\mathbf{a} = (a_1, \ldots, a_m)$ and a vector of length m, $\mathbf{x} = (x^1, \ldots, x^m)^T$, the multiplication is denoted as

$$\mathbf{ax} = (a_1, \ldots, a_m) \begin{pmatrix} x_1 \\ \cdots \\ x_m \end{pmatrix} = \sum_i a_i x^i . \qquad (5.1)$$

Notice that there is no "dot" in Eq. (5.1)—it is not a scalar product but a pairing between covectors and vectors.

© The Author(s), under exclusive license to Springer Nature Switzerland AG 2025
V. Putkaradze, *A Concise Introduction to Classical Mechanics*,
Surveys and Tutorials in the Applied Mathematical Sciences 16,
https://doi.org/10.1007/978-3-031-84977-0_5

Now, matrix multiplication between $n \times m$ matrices and m-vectors is defined from the point of view that a matrix is a vector of row vectors, so we write

$$\mathbb{A}\mathbf{x} = \begin{pmatrix} \mathbf{a}_1 \\ \dots \\ \mathbf{a}_m \end{pmatrix} \begin{pmatrix} x_1 \\ \dots \\ x_m \end{pmatrix} = \begin{pmatrix} \mathbf{a}_1\mathbf{x} \\ \dots \\ \mathbf{a}_m\mathbf{x} \end{pmatrix} = a^i_j x^j \quad \text{(sum over } j\text{)}. \tag{5.2}$$

The result of a product of an $n \times m$ matrix and an m-vector is an n-vector.

Next, to compute matrix multiplication between $n \times m$ matrix $\mathbb{A}$ and $m \times l$ matrix $\mathbb{B}$, we view the matrix $\mathbb{A}$ as the n-vector of rows of dimension m, and $\mathbb{B}$ as a row-vectors of columns of dimension l. Then, the (i, j)-th component of the product is given by

$$(\mathbb{A}\mathbb{B})^i_j = a^i_k b^k_j \quad \text{(sum over } k\text{)}. \tag{5.3}$$

Note that the second dimension of $\mathbb{A}$ and the first dimension of $\mathbb{B}$ have to match; otherwise, the product $\mathbb{A}\mathbb{B}$ is not defined.

Matrix Commutator As we have seen, in general, for an $n \times m$ matrix $\mathbb{A}$ and $m \times l$ matrix $\mathbb{B}$, the product $\mathbb{B}\mathbb{A}$ is not defined if $l \neq n$, even though $\mathbb{A}\mathbb{B}$ is defined. If $l = n$, the result $\mathbb{B}\mathbb{A}$ is defined and is $m \times m$ matrix, whereas $\mathbb{A}\mathbb{B}$ is an $n \times n$ matrix. When $\mathbb{A}$ and $\mathbb{B}$ are both square matrices of the same dimension $m = n = l$, then $\mathbb{A}\mathbb{B}$ and $\mathbb{B}\mathbb{A}$ have the same dimension. In general, $\mathbb{A}\mathbb{B} \neq \mathbb{B}\mathbb{A}$, and it makes sense to define the *commutator*, which is also an $n \times n$ matrix:

$$[\mathbb{A}, \mathbb{B}] = \mathbb{A}\mathbb{B} - \mathbb{B}\mathbb{A}. \tag{5.4}$$

Eigenvalues/Eigenvectors For a square matrix $\mathbb{A}$, a vector $\mathbf{v} \neq 0$ and a scalar λ such that $\mathbb{A}\mathbf{v} = \lambda\mathbf{v}$ are called eigenvectors and eigenvalues, respectively. A real symmetric $n \times n$ matrix has only real eigenvalues; however, they do not necessarily have to be distinct.

Matrix Determinant The determinant of an $n \times n$ matrix $\det\mathbb{A}$ is defined recursively as

$$\mathbb{A} \text{ is } 2 \times 2 : \det\mathbb{A} = a^1_1 a^2_2 - a^1_2 a^2_1$$

$$\mathbb{A} \text{ is } n \times n : \det\mathbb{A} = \sum_j (-1)^{i+j} a^i_j \det\mathbb{M}_{i,j}, \quad \text{(no sum over } i\text{) where} \tag{5.5}$$

$$\mathbb{M}_{i,j} \text{ is } (n-1) \times (n-1) : \mathbb{A} \text{ without } i\text{-th column and } j\text{-th row}.$$

For completeness, one defines the determinant of a scalar (1×1 matrix) as simply that scalar: $\det a = a$.

Solutions of Linear Equations, Eigenvalues, and Eigenvectors If $\mathbb{A}$ is an $n \times n$ matrix, $\mathbf{x}$ and $\mathbf{f}$ are n-vectors, then $\mathbb{A}\mathbf{x} = \mathbf{f}$ is called a system of linear equations. The system has a unique solution if and only if $\det \mathbb{A} \neq 0$.

We can define the $n \times n$ unity matrix Id as having 1 on diagonals and 0 off-diagonal. The matrix is called unity matrix since for any $n \times n$ matrix $\mathbb{A}$, it satisfies $\mathbb{A} \,\mathrm{Id} = \mathrm{Id}\, \mathbb{A} = \mathbb{A}$. Sometimes we will put an explicit size of Id in the notation, for example, $\mathrm{Id}_{3 \times 3}$.

We rewrite the eigenvalue condition as $(\mathbb{A} - \lambda \mathrm{Id})\mathbf{v} = \mathbf{0}$. If the determinant of the matrix in parentheses is not zero, then the only solution is $\mathbf{v} = \mathbf{0}$. Thus, the eigenvalue condition for the matrix $\mathbb{A}$, i.e., for $\mathbf{v} \neq \mathbf{0}$ to exist, is

$$\det(\mathbb{A} - \lambda \mathrm{Id}) = 0. \tag{5.6}$$

The condition (Eq. (5.6)) produces an n-order polynomial in λ, called characteristic polynomial for $\mathbb{A}$. Thus, there are exactly n eigenvalues—the roots of characteristic polynomial—in the complex plane, including the multiplicity.

Inverse Matrix If $\det \mathbb{A} \neq 0$, then there exists a matrix $\mathbb{A}^{-1}$ such that $\mathbb{A}^{-1}\mathbb{A} = \mathbb{A}\mathbb{A}^{-1} = \mathrm{Id}$. That matrix $\mathbb{A}^{-1}$ is called the *inverse matrix* of $\mathbb{A}$.

Remark 5.1 (Nondegenerate Lagrangians) In what follows, we will always assume that the Hessian $\mathbb{H}$ of the Lagrangian is non-singular:

$$H_{\alpha\beta} = \frac{\partial^2 L}{\partial \dot{q}^\alpha \partial \dot{q}^\beta}, \quad \det \mathbb{H} \neq 0. \tag{5.7}$$

This requirement is needed to avoid unpleasant problems. For example, if you try to write Euler-Lagrange equations for the Lagrangian $L = \mathbf{a}(\mathbf{q}) \cdot \dot{\mathbf{q}} - U(\mathbf{q})$, where $\mathbb{H}$ vanishes identically, you get first-order Euler-Lagrange equations that may or may not have solutions depending on the functions $\mathbf{a}(\mathbf{q})$ and $U(\mathbf{q})$. In general, if the Hessian $\mathbb{H}(\mathbf{q}, \dot{\mathbf{q}})$ in Eq. (5.7) vanishes on some surfaces in the phase space, the computation of Euler-Lagrange equations around these surfaces becomes tricky and is definitely beyond the scope of this book.

Useful Tricks for Finding Eigenvalues For diagonal, upper triagonal (entries only on diagonal and above), or lower triagonal (entries only on diagonal and below) matrices, the eigenvalues are simply equal to the values of the diagonal entries. If a matrix can be separated into square "blocks" (sub-matrices), i.e., there are square pieces of nonzero elements about the diagonal, then one can just look for eigenvalues for these smaller sub-matrices.

5.2   Linear Stability of a Mechanical System

There is a general method to compute the small vibrations about equilibrium for
an arbitrary system as long as the steady states are known. Suppose Q is an n-
dimensional configuration manifold, and q_0 is the equilibrium (fixed) point of the
mechanical system. This means that $\mathbf{q}(t) = \mathbf{q}_0$ for all times, and $(\mathbf{q} = \mathbf{0}, \dot{\mathbf{q}} = \mathbf{0})$
is a solution of Euler-Lagrange equations. If $L = T(\mathbf{q}, \dot{\mathbf{q}}) - V(\mathbf{q})$, then for this to
happen, we need $\frac{\partial V}{\partial \mathbf{q}} = \mathbf{0}$ when $\mathbf{q} = \mathbf{q}_0$, assuming $T(\mathbf{q}, \dot{\mathbf{q}})$ is a quadratic function of
$\dot{\mathbf{q}}$.

The next step is to assume $\mathbf{q} = \mathbf{q}_0 + \epsilon \mathbf{x}$, $\dot{\mathbf{q}} = \epsilon \dot{\mathbf{x}}$ and expand the Lagrangian to the
second order in ϵ. Most mechanical systems have kinetic energies that are quadratic
in velocities, so we perform the calculation for these systems. A more general case
can be considered by expanding the Lagrangian to the second order in $\dot{\mathbf{q}}$ around the
equilibrium. Then, the kinetic energy is approximated as

$$T = \frac{1}{2}\dot{\mathbf{q}}^T \mathbb{A}(\mathbf{q})\dot{\mathbf{q}} = \frac{1}{2}A_{\alpha\beta}(\mathbf{q})\dot{q}^\alpha \dot{q}^\beta = \frac{1}{2}\epsilon^2 \dot{\mathbf{x}}^T \mathbb{A}(\mathbf{q}_0)\dot{\mathbf{x}} + O(\epsilon^3)$$

$$= \frac{1}{2}\epsilon^2 A_{\alpha\beta}(\mathbf{q}_0)\dot{x}^\alpha \dot{x}^\beta + O(\epsilon^3) := \frac{1}{2}\epsilon^2 \dot{\mathbf{x}}^T \mathbb{M}\dot{\mathbf{x}} + O(\epsilon^3), \quad \mathbb{M} := \mathbb{A}(\mathbf{q}_0).$$

$$(5.8)$$

The physical meaning of the matrix $\mathbb{M}$ is the mass matrix computed at the
equilibrium point. The matrix $\mathbb{M}$ is symmetric since the matrix $\mathbb{A}$ is symmetric.
For the potential energy, we expand $V(\mathbf{q})$ about $\mathbf{q} = \mathbf{q}_0$ up to the second order in ϵ:

$$V(\mathbf{q}) = V(\mathbf{q}_0) + \epsilon \underbrace{\frac{\partial V}{\partial \mathbf{q}}(\mathbf{q}_0) \cdot \mathbf{x}}_{=0} + \frac{1}{2}\epsilon^2 \frac{\partial^2 V}{\partial q^\alpha \partial q^\beta}(\mathbf{q}_0)x^\alpha x^\beta + O(\epsilon^3)$$

$$(5.9)$$

$$= V_0 + \frac{1}{2}\epsilon^2 K_{\alpha\beta}x^\alpha x^\beta + O(\epsilon^3) = V_0 + \frac{1}{2}\epsilon^2 \mathbf{x}^T \mathbb{K}\mathbf{x} + O(\epsilon^3).$$

Here, $\mathbb{K}$ is the Hessian matrix of second partial derivatives of the potential, evaluated
at the equilibrium point, also known as the stiffness matrix. The α, β component of
$\mathbb{K}$ is $K_{\alpha\beta}$. Naturally, $\mathbb{K}$ is symmetric. Then, to the second order in ϵ, the Lagrangian
is given as

$$L = -V_0 + \frac{1}{2}\epsilon^2 \left(\dot{\mathbf{x}}^T \mathbb{M}\dot{\mathbf{x}} - \mathbf{x}^T \mathbb{K}\mathbf{x}\right) + O(\epsilon)^3$$

$$(5.10)$$

$$= -V_0 + \frac{1}{2}\epsilon^2 \left(M_{\alpha\beta}\dot{x}^\alpha \dot{x}^\beta - K_{\alpha\beta}x^\alpha x^\beta\right) + O(\epsilon^3).$$

The Euler-Lagrange equations are then given as

$$M_{\alpha\beta}\ddot{x}^\beta + K_{\alpha\beta}x^\beta = 0, \quad \text{in vector form} \quad \mathbb{M}\ddot{\mathbf{x}} + \mathbb{K}\mathbf{x} = \mathbf{0}. \quad (5.11)$$

Equation (5.11) is a linear, second-order differential equation with constant coefficients. We assume the exponential dependence $\mathbf{x} = e^{i\omega t}\mathbf{v} = (\cos \omega t + i \sin \omega t)\mathbf{v}$. (Alternatively, we can assume $\mathbf{x} = \mathbf{v}e^{\alpha t}$ and then use $\alpha = i\omega$ in equations since we will be interested in oscillations). We obtain a condition for the frequency ω:

$$\mathbb{K}\mathbf{v} = \omega^2 \mathbb{M}\mathbf{v}, \quad \text{or} \quad \left(\mathbb{K} - \omega^2 \mathbb{M}\right)\mathbf{v} = \mathbf{0}. \tag{5.12}$$

Equation (5.12) is called the generalized eigenvalue problem of matrix $\mathbb{K}$ with respect to matrix $\mathbb{M}$. A nontrivial solution for eigenvalues exists if and only if the characteristic equation for the generalized eigenvalues ω^2 is satisfied:

$$\det\left(\mathbb{K} - \omega^2 \mathbb{M}\right) = 0. \tag{5.13}$$

To prove that eigenvalues ω^2 defined by Eq. (5.13) are real, we recall a useful theorem from linear algebra. Consider two quadratic forms: the kinetic energy $\frac{1}{2}\mathbb{M}\dot{\mathbf{x}} \cdot \dot{\mathbf{x}}$ and the potential energy $\frac{1}{2}\mathbb{K}\mathbf{x} \cdot \mathbf{x}$, with the matrix $\mathbb{M}$ defining the first quadratic form being symmetric and positive definite. Then, from the theorem on diagonalization of quadratic forms, there exists a transformation to coordinates $\mathbf{X} = \mathbb{C}\mathbf{x}$ such that

$$\mathbb{M}\dot{\mathbf{x}} \cdot \dot{\mathbf{x}} \to |\dot{\mathbf{X}}|^2 = \sum_{\alpha=1}^{n} \dot{X}_\alpha^2, \quad \mathbb{K}\mathbf{x} \cdot \mathbf{x} \to \sum_{\alpha=1}^{n} \lambda_\alpha X_\alpha^2, \tag{5.14}$$

for some real numbers λ_α, $\alpha = 1, \ldots, n$; see [1], Chap. 23. The dynamics of each component α of $\mathbf{X}$ is then given by

$$\ddot{X}^\alpha + \lambda_\alpha X^\alpha = 0, \quad \alpha = 1, \ldots, n \quad \mathbf{x} = \mathbb{C}^{-1}\mathbf{X}. \tag{5.15}$$

Depending on the sign of λ_α in Eq. (5.15), the solution for X^α will be

$$\lambda > 0 : \omega_\alpha = \sqrt{\lambda_\alpha}, \quad X^\alpha = C \cos \omega_\alpha t + D \sin \omega_\alpha t,$$
$$\lambda = 0 : \omega_\alpha = 0, \quad X^\alpha = Ct + D, \tag{5.16}$$
$$\lambda < 0 : \kappa_\alpha = \sqrt{-\lambda_\alpha}, \quad X^\alpha = C \cosh \kappa_\alpha t + D \sinh \kappa_\alpha t,$$

and $\mathbf{x}$ is obtained as a linear combination of solutions given by Eq. (5.16) multiplied by eigenvectors $\mathbf{v}_\alpha$. We can write it in a compact form:

$$\mathbf{x} = \sum_{\alpha : \omega_\alpha \neq 0} \mathbf{v}^\alpha (C_\alpha \cos \omega_\alpha t + D_\alpha \sin \omega_\alpha t) + \sum_{\alpha : \omega_\alpha = 0} \mathbf{v}^\alpha (C_\alpha + D_\alpha t), \tag{5.17}$$

remembering that for ω_α being purely imaginary $\omega_\alpha = i\kappa_\alpha$, we have $\cos \omega_\alpha t = \cosh \kappa_\alpha t$ and $\sin \omega_\alpha t = i \sinh \kappa_\alpha t$. When repeated eigenvalues are present, the solutions of a *general* higher-dimensional linear equations can become quite a bit more complicated; see [34], Chap. 6. Fortunately, for the case of the diagonalizable kinetic and potential energies as in Eq. (5.14), the general solution can still be written down according to Eq. (5.15), and thus Eq. (5.17) still holds. For example, some linear systems with repeated imaginary eigenvalues $\pm i\omega_\alpha$ have resonance-like terms in the solutions that grow as $t \sin \omega_\alpha t$; however, for the mechanical system of the type we consider here, such solutions do not occur.

There is another way of solving the problem by bringing the system (Eq. (5.12)) to the standard eigenvalue problem. We know that the matrix $\mathbb{M}$ is nondegenerate since the Lagrangian is nondegenerate. Denote $\lambda = \omega^2$ and multiply (Eq. (5.12)) by $\mathbb{M}^{-1}$ on the left. Denote $\Lambda = \mathbb{M}^{-1}\mathbb{K}$ and obtain the equation for eigenvalues of Λ:

$$\Lambda \mathbf{v} = \lambda \mathbf{v} \quad \Rightarrow \quad \det(\Lambda - \lambda \mathrm{Id}) = 0, \quad \Lambda = \mathbb{M}^{-1}\mathbb{K}, \quad \lambda = \omega^2. \tag{5.18}$$

Eigenvalues of Λ are the same as obtained from the solution of Eq. (5.12). As we proved above, these eigenvalues are real. The eigenvectors of Λ also coincide exactly with the generalized eigenvectors given by Eq. (5.12), and the general solution is also given by Eq. (5.17). Thus, these two methods are equivalent.

If all ω_α are real, then the linearized solution (Eq. (5.17)) does not grow exponentially. This case corresponds to all eigenvalues λ_α being nonnegative. We call such systems *linearly stable*. The term proportional to t in Eq. (5.17) corresponds to a free shift along a variable with $\lambda_\alpha = 0$ and does not represent an instability for the purpose of this discussion. The constants C_α and D_α are obtained from the initial conditions, which are usually given as $\mathbf{x}(0) = \mathbf{x}_0$, $\dot{\mathbf{x}}(0) = \mathbf{w}_0$. Define the matrix of eigenvectors $\mathbb{V} = (\mathbf{v}_1, \ldots, \mathbf{v}_n)$, which we assume is invertible. Then, these coefficients C_α and D_α can be found from the following linear equations:

$$C_\alpha = \left(\mathbb{V}^{-1}\mathbf{x}_0\right)_\alpha, \quad \mathbf{D} = \begin{cases} \dfrac{1}{\omega_\alpha}\left(\mathbb{V}^{-1}\mathbf{w}_0\right)_\alpha, & \omega_\alpha \neq 0 \\[2mm] \left(\mathbb{V}^{-1}\mathbf{w}_0\right)_\alpha, & \omega_\alpha = 0. \end{cases} \tag{5.19}$$

In practice, to find, *e.g.*, $\mathbf{c} = \mathbb{V}^{-1}\mathbf{x}_0$, you would solve the equation $\mathbb{V}\mathbf{c} = \mathbf{x}_0$. In order to find D_α, you would first find a vector $\mathbf{W}$ from the linear system $\mathbb{V}\mathbf{W} = \mathbf{w}_0$ and then find $D_\alpha = -W_\alpha/\omega_\alpha$. It is, as a rule, easier to solve linear equations than to invert matrices. The case of general ω_α including zero eigenvalues is considered similarly.

On the other hand, if at least one eigenvalue $\lambda_\alpha = \omega_\alpha^2$ in Eqs. (5.12) or (5.18) is negative, this mode experiences exponential growth in time since the corresponding $\omega_\alpha = \pm i\alpha$ will be purely imaginary, and thus there will be terms $e^{\pm \alpha t}$ instead of oscillating terms in Eq. (5.17). These terms correspond to the linear instability of the system.

On the Standard and Generalized Eigenvalue Approaches Computation of the eigenfrequencies through the generalized eigenvalue problem (Eq. (5.13)) tends to be simpler than the computation of the eigenvalues through the eigenvalues of the matrix Λ, since Eq. (5.13) does not require inverting the matrix $\mathbb{M}$. However, utilizing generalized eigenvalues requires the use of some material not commonly studied in linear algebra courses. Moreover, if the matrix $\mathbb{M}$ is easily invertible, which is the case for most examples considered in this book, computing Λ is advantageous since it readily shows the meaning of the combinations of several physical parameters. We will mostly use the approach relying on standard eigenvalues of the matrix Λ as being the most familiar to the reader. Moreover, eigenvalues of Λ have a precise physical meaning and can be rescaled to remove potentially very large or very small scaling factors, which can be harder to see in the direct application of Eq. (5.13). An example of such useful scaling is presented in the computations of two- and three-atom molecules in Examples 5.3 and 5.4. An interested reader may redo the examples below using the application of Eq. (5.13) to compute the generalized eigenvalues and Eq. (5.12) to find the generalized eigenvectors.

Linear Versus Nonlinear Stability Analyzing the fully nonlinear stability of the full system is substantially more complicated than presented here. The big question is whether the nonlinear system behaves similar to that of the linearized system. Indeed, one can prove that if there is a linear *instability*, the fully nonlinear system will also be unstable. More generally, if none of the eigenvalues have real part 0, then there is a local correspondence between the behaviors of the nonlinear and linearized system. However, if the real part of eigenvalues for a linearized system is exactly zero, then nothing can be said in general. See [34], Chap. 8, for a popular and clear exposition of that fact. We will see that the case we call "linearly stable" will actually have all real parts of eigenvalues equal to zero. Thus, nonlinear systems may be stable or unstable; no statement is made on the stability of the nonlinear systems here.

5.3 Examples

Let us first consider a few examples from Sect. 3.7.

Example 5.1 (From Sect. 3.7: A Single Pendulum) The Lagrangian is given by Eq. (3.18). Equilibrium points are given by setting $\frac{\partial L}{\partial \varphi} = 0$, yielding $\sin \varphi_* = 0$, or $\varphi_* = 0$ and $\varphi_* = \pi$. The first equilibrium point corresponds to the pendulum pointing down, and the other one to the pendulum pointing up. We can conjecture that the first one is stable and the second one is unstable.

Let us first consider the critical point $\varphi_* = 0$ Approximating the Lagrangian from Eq. (3.18) for small angles φ, we get

$$L \simeq \frac{1}{2}ml^2\dot{\varphi}^2 - \frac{1}{2}mgl\varphi^2 + \text{const}. \tag{5.20}$$

The mass matrix is one dimensional $\mathbb{M} = ml^2$, and the matrix $\mathbb{K} = mgl$ is also one dimensional. Thus, $\Lambda = \mathbb{M}^{-1}\mathbb{K} = g/l = \omega_0^2$. There is clearly only one eigenvalue $\lambda = \omega_0^2$. Small oscillations have the frequency $\omega_0 = \sqrt{g/l}$ as expected. Similarly, for the equilibrium point $\varphi_* = \pi$, the mass matrix $\mathbb{M}$ is the same, but $\mathbb{K} = -mgl$. Then, $\lambda = -\omega_0^2$, so the equilibrium point is unstable.

Example 5.2 (From Sect. 3.7: An Elastic Pendulum) The Lagrangian for the elastic pendulum is given by Eq. (3.22). Before we proceed to linear stability, let us find the equilibrium points, which are given by $\frac{\partial L}{\partial \varphi} = 0$ and $\frac{\partial L}{\partial l} = 0$. These conditions, defining the equilibrium states (l_*, φ_*), lead to

$$\sin \varphi_* = 0, \quad U'(l_*) = 0. \tag{5.21}$$

Physically, these conditions correspond to the pendulum hanging up or down while the stretch of the elastic spring compensates for the action of gravity on the bob. We will only consider the case $(l = l_*, \varphi_* = 0)$ (pendulum down). The equilibrium point $(l = l_*, \varphi_* = \pi)$ can be considered in the same way, and that equilibrium point will, of course, be unstable.

Let us not turn to the actual calculation. The Lagrangian (Eq. (3.22)), expanded for small oscillations about the equilibrium $(l = l_*, \varphi_* = 0)$, by defining $x = l - l_* \ll l$, $\varphi \ll 1$ is given

$$L \simeq T - V = \frac{1}{2}m \left(\dot{x}^2 + l_*^2 \dot{\varphi}^2 \right) - mgl_* \frac{\varphi^2}{2} - \frac{1}{2}U''(l_*)x^2. \tag{5.22}$$

Note that, for example, the term $l^2\dot{\varphi}^2$ is approximated as $l_*^2\dot{\varphi}^2$ since the terms proportional to $x\varphi^2$ are of lower order. Let us, for shortness, define $\kappa := U''(l_*)$, which is the effective stiffness of the spring at equilibrium. The matrices are given by

$$\mathbb{M} = m \begin{pmatrix} 1 & 0 \\ 0 & l_*^2 \end{pmatrix}, \quad \mathbb{K} = \begin{pmatrix} \kappa & 0 \\ 0 & mgl_* \end{pmatrix}, \quad \Lambda = \mathbb{M}^{-1}\mathbb{K} = \begin{pmatrix} \Omega_0^2 & 0 \\ 0 & \omega_0^2 \end{pmatrix}, \tag{5.23}$$

where we defined two frequencies: that of the elastic spring $\Omega_0^2 = \kappa/m$ and that of the pendulum $\omega_0^2 = g/l$. Eigenvalues of Λ from Eq. (5.23) are (Ω_0^2, ω_0^2) and eigenfrequencies are Ω_0 and ω_0. Eigenvectors are $V_1 = (1, 0)^T$ for $\lambda = \Omega_0^2$ and $V_2 = (0, 1)^T$ for $\lambda = \omega_0^2$. The evolution of the system according to the linearized equations is given by

$$\begin{pmatrix} x \\ \varphi \end{pmatrix} = \begin{pmatrix} 1 \\ 0 \end{pmatrix} (C_1 \cos \Omega_0 t + D_1 \sin \Omega_0 t) + \begin{pmatrix} 0 \\ 1 \end{pmatrix} (C_2 \cos \omega_0 t + D_2 \sin \omega_0 t)$$

$$= \begin{pmatrix} C_1 \cos \Omega_0 t + D_1 \sin \Omega_0 t \\ C_2 \cos \omega_0 t + D_2 \sin \omega_0 t \end{pmatrix}.$$

$$\tag{5.24}$$

Alternatively, we could compute the eigenvalues directly from the equation (Eq. (5.12)):

$$|\mathbb{K} - \lambda \mathbb{M}| = \begin{vmatrix} \kappa - ml_*^2 \lambda & 0 \\ 0 & mgl_* - ml_*^2 \lambda \end{vmatrix} \Rightarrow \lambda_1 = \frac{\kappa}{m} = \Omega_0^2, \ \lambda_2 = \frac{g}{l_*} = \omega_0^2.$$

$$(5.25)$$

The eigenfrequencies are Ω_0 and ω_0, and the generalized eigenvectors are still $(1, 0)^T$ and $(0, 1)^T$, with the solution still given by (5.24).

You should now be able to perform the linear stability analysis for the rest of the examples (Examples 3.3, 3.4, and 3.5) from Sect. 3.7. We present two additional examples below, demonstrating what happens when the matrix Λ has zero eigenvalues.

Example 5.3 (A Two-Atom Molecule: 1D Motion) Suppose two atoms of masses m_1 and m_2 are moving on a straight line, so $q_1 \in \mathbb{R}$ and $q_2 \in \mathbb{R}$. The potential connects these atoms, which depends only on the difference between $V(q_1, q_2) = F(q_1 - q_2)$. Suppose that $q_1 - q_2 = l$ is the equilibrium distance, so $F'(l) = 0$. Let us denote the effective spring constant of $F''(l) = k$.

The new variables are $x_1 = q_1$, $x_2 = q_2 \pm l$, where we choose the $+$ sign if the second particle is to the left of the first one and the minus sign otherwise. In fact, we can take $x_1 = q_1 + a$, $x_2 = l + a + q_2$ for any $a \in \mathbb{R}$. That ambiguity in choosing the origin translates, later, to the existence of zero eigenvalues. Note that close to the equilibrium, $V \simeq \frac{k}{2}(x_1 - x_2)^2$. The matrices $\mathbb{M}$ and $\mathbb{K}$ are given by

$$\mathbb{M} = \begin{pmatrix} m_1 & 0 \\ 0 & m_2 \end{pmatrix}, \quad \mathbb{K} = k \begin{pmatrix} 1 & -1 \\ -1 & 1 \end{pmatrix}. \tag{5.26}$$

In what follows, it is helpful to define the natural frequency for one of the masses $\omega_0^2 = k/m_1$ and mass ratio $\mu = m_1/m_2$. The matrix defining the eigenvalues Λ is computed as

$$\Lambda = \mathbb{M}^{-1}\mathbb{K} = \omega_0^2 \begin{pmatrix} 1 & -1 \\ -\mu & \mu \end{pmatrix}. \tag{5.27}$$

For simplicity, let us compute the dimensionless eigenvalues of the matrix Λ/ω_0^2 and then simply multiply the found eigenvalues by ω_0^2. This corresponds to finding the frequencies in terms of units ω_0 and is useful if you trying to find molecular frequencies—it is much easier to deal with quantities of order 1 than $10^{13} - 10^{14}$ Hz. The eigenvalues of Λ/ω_0^2 are computed as follows:

$$\begin{vmatrix} 1 - \lambda & -1 \\ -\mu & \mu - \lambda \end{vmatrix} = \lambda^2 - (\mu + 1)\lambda = 0, \quad \Rightarrow \quad \lambda_1 = 0, \quad \lambda_2 = \mu + 1. \tag{5.28}$$

The eigenvector for the eigenvalue λ_i, called $\mathbf{v}_i$, is given by $(\Lambda/\omega^2 - \lambda_i)\mathbf{v}_i = \mathbf{0}$. When substituting λ_i in the matrix, the rows must become linearly dependent. We get, for $\lambda_1 = 0$, and $\mathbf{v}_1 = (w_1, w_2)$,

$$\begin{pmatrix} 1 & -1 \\ -\mu & \mu \end{pmatrix} \begin{pmatrix} w_1 \\ w_2 \end{pmatrix} = \begin{pmatrix} 0 \\ 0 \end{pmatrix}, \quad \Rightarrow w_1 - w_2 = 0 \quad , w_1 = w_2 \tag{5.29}$$

and we can take the eigenvector $\mathbf{v}_1 = (1, 1)^T$. Sometimes, one normalizes the eigenvectors by dividing $\mathbf{v}_1$ by its norm, but it is not necessary here.

Similarly, for $\lambda_2 = \mu + 1$, repeating Eq. (5.29) gives $\mathbf{v}_2 = (1, -\mu)^T$. The general solution for the motion of the atoms about the equilibrium is

$$\begin{pmatrix} x_1 \\ x_2 \end{pmatrix} = \underbrace{\begin{pmatrix} 1 \\ 1 \end{pmatrix} (C_1 + D_1 t)}_{\text{Uniform motion}} + \begin{pmatrix} 1 \\ -\mu \end{pmatrix} \left(C_2 \cos \sqrt{\mu + 1} \omega_0 t + D_2 \sin \sqrt{\mu + 1} \omega_0 t \right).$$

$$\tag{5.30}$$

The first term, corresponding to $\lambda_1 = 0$, describes the uniform motion of both atoms with the same speed. This motion can be removed by going to a proper inertial coordinate frame and thus is not considered a part of oscillatory motion. There is only one frequency of motion $\omega = \sqrt{\mu + 1} \omega_0$ for this system.

Example 5.4 (A Three-Atom Linear Molecule) Suppose there is a three-unit molecule that has two distinct atoms, one of mass M in the center, at $q_2 = 0$, and two of masses m on the left and the right side $q_1 = -l$ and $q_3 = l$. We need to make some assumptions about the potential energy, namely, that only the neighboring atoms interact; otherwise, the system will become quite complex. On the other hand, the assumption of nearest-neighbor interaction is physical and widely used in practice. Assuming that the potential of interaction depends only on the relative distance, as in the previous case, we write the Lagrangian as

$$L = \frac{1}{2} \left(m\dot{q}_1^2 + m\dot{q}_3^2 + M\dot{q}_2^2 \right) - F(q_1 - q_2) - F(q_2 - q_3). \tag{5.31}$$

We can take $x_1 = q_1 + l$, $x_2 = q_2$, and $x_3 = q_3 - l$. The approximate Lagrangian (dropping the constant term $-2F_0$) is

$$L = \frac{1}{2} \left(m\dot{x}_1^2 + m\dot{x}_3^2 + M\dot{x}_2^2 \right) - \frac{k}{2} \left((x_1 - x_2)^2 + (x_3 - x_2)^2 \right). \tag{5.32}$$

We shall just present the results as the solution is very similar to the two-molecule case. Denoting, again, $\mu = m/M$, we define the mass $\mathbb{M}$, stiffness $\mathbb{K}$, and Λ matrices as

$$\mathbb{M} = \begin{pmatrix} m & 0 & 0 \\ 0 & M & 0 \\ 0 & 0 & m \end{pmatrix}, \; \mathbb{K} = k \begin{pmatrix} 1 & -1 & 0 \\ -1 & 2 & -1 \\ 0 & -1 & 1 \end{pmatrix}, \; \Lambda = \omega_0^2 \begin{pmatrix} 1 & -1 & 0 \\ -\mu & 2\mu & -\mu \\ 0 & -1 & 1 \end{pmatrix}. \tag{5.33}$$

The eigenvalues of Λ/ω_0^2 are $\lambda_1 = 0$, $\lambda_2 = 1$, and $\lambda_2 = 2\mu + 1$. The corresponding eigenvectors are $\mathbf{v}_1 = (1, 1, 1)^T$, $\mathbf{v}_2 = (1, 0, -1)^T$, and $\mathbf{v}_3 = (1, -2\mu, 1)^T$. (Prove both of these statements!) Again, λ_0 with the eigenvector $\mathbf{v}_0$ corresponds to the translational motion of the whole system and is not related to any oscillations. The eigenvalue λ_2 with eigenvector $\mathbf{v}_2$ corresponds to the central atom being at rest and two side atoms are moving in opposite directions, with the angular frequency ω_0. Finally, the third eigenvalue λ_3 corresponds to the combined motion of all atoms with the angular frequency $\omega = \omega_0\sqrt{2\mu + 1}$.

5.4 Looking Forward

The question of linear stability considered here forms the foundation of local analysis of mechanical systems. As we mentioned, the linear stability considered here, in general, doesn't prove the stability of the instability of the nonlinear systems. The question of the stability of Lagrangian and Hamiltonian systems (see the next chapter) is actually very difficult because of the absence of dissipation. Some of the nonlinear stability methods are Lyapunov's method and Lasalle's invariance principle [34], and many other methods have been developed particularly for applications in mechanics [35, 36]. Of particular note are the applications of Casimir-energy methods and other fully nonlinear methods [37–40]. Various methods of fully nonlinear analysis have been developed for fluids, plasmas, and other applications [41–44]. An interesting and somewhat counterintuitive recent discovery is that dissipation does not necessarily lead to stability; in contrast, some systems exhibit instabilities due to the introduction of dissipation [45–47].

Thus, while the exposition of linear stability is quite standard and well-established, the *nonlinear* stability for mechanical systems is a highly active research area, often requiring individual approaches to each particular problem.

5.5 Practice Problems

1. Figure 5.1a (Left): A mass m is attached with four massless springs, each having the equilibrium length l, to four unmovable blocks. The blocks are set at the points $(0, \pm l)$, $(\pm l, 0)$. The springs have a stiffness of c and μc, as illustrated.

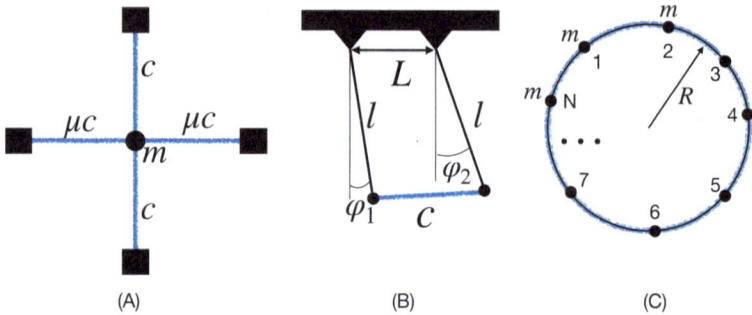

Fig. 5.1 Setup for the problems in this section

Find the frequencies of small oscillations around the equilibrium $(0, 0)$ and compute the evolution of the mass. Neglect all gravity forces.

2. Figure 5.1b (Center): Two pendulums of length l are separated by the distance L and are also connected with a massless spring of equilibrium length L and stiffness c. Find the frequencies of small oscillations and compute the evolution of the solutions about the equilibrium. The masses of the pendulum's bobs are m.

3. Figure 5.1c (Right): N equal masses sliding without friction along a circle of radius R. The masses are connected by massless springs of stiffness c and equal equilibrium length $2\pi R/N$. Hence, the equilibrium is at the state when all the masses are positioned along the circle at equal distances. Find the frequencies of small oscillations about the equilibrium.

Hint. Look for (possibly complex) eigenvectors of the form $\mathbf{v} = (1, q, q^2, \ldots, q^{n-1})^T$ for some q and then bring the system to real solutions.

Chapter 6
Hamiltonian Systems

6.1 Derivations of Hamilton's Equations

Let us go back to our derivation of the Euler-Lagrange equations (Eq. (2.11)). In the absence of external forces $Q_\alpha = 0$, we can write these equations expressing the rate of change of momentum:

$$\dot{\mathbf{p}} = \frac{\partial L}{\partial \mathbf{q}}, \quad \text{where} \quad \mathbf{p} = \frac{\partial L}{\partial \dot{\mathbf{q}}} \quad \text{is the momentum}. \tag{6.1}$$

This equation is, however, still a second-order equation in $\mathbf{q}$, since $\mathbf{p} = \mathbf{p}(\dot{\mathbf{q}}, \mathbf{q}, t)$. Our goal is to use $(\mathbf{p}, \mathbf{q})$ as the new variables, instead of $(\mathbf{q}, \dot{\mathbf{q}})$. We thus solve $\mathbf{p} = \mathbf{p}(\dot{\mathbf{q}}, \mathbf{q}, t)$ for $\dot{\mathbf{q}}$ and express $\dot{\mathbf{q}} = \dot{\mathbf{q}}(\mathbf{q}, \mathbf{p}, t)$. This transformation is possible if the Lagrangian is nondegenerate, as we assumed before in Eq. (5.7). Let us show how to perform this transformation on a particular example. If we assume the Lagrangian of the form

$$L(\mathbf{q}, \dot{\mathbf{q}}) = \frac{1}{2} M_{\alpha\beta}(\mathbf{q}) \dot{q}^\alpha \dot{q}^\beta - V(\mathbf{q}) = \frac{1}{2} \dot{\mathbf{q}} \cdot \mathbb{M}(\mathbf{q}) \dot{\mathbf{q}} - V(\mathbf{q}). \tag{6.2}$$

Then, if we assume that the mass matrix $\mathbb{M}$ is invertible, by (5.7), the velocities $\dot{\mathbf{q}}$ are expressed in terms of coordinates and momenta $(\mathbf{q}, \mathbf{p})$ according to the following calculation:

$$\mathbf{p} = \mathbb{M}(\mathbf{q}) \dot{\mathbf{q}}, \quad \text{in coordinates: } p_\beta = M_{\alpha\beta}(\mathbf{q}) \dot{q}^\beta,$$

$$\dot{\mathbf{q}} = \mathbb{M}(\mathbf{q})^{-1} \mathbf{p}, \quad \text{in coordinates: } \dot{q}^\alpha = \left(\mathbb{M}(\mathbf{q})^{-1} \right)^{\alpha\beta} p_\beta. \tag{6.3}$$

The solution (Eq. (6.3)) is a particular case of the dependence $\dot{\mathbf{q}} = \dot{\mathbf{q}}(\mathbf{q}, \mathbf{p}, t)$ for the Lagrangian that is quadratic in velocities. In what follows, we will not assume any

V. Putkaradze, *A Concise Introduction to Classical Mechanics*,
Surveys and Tutorials in the Applied Mathematical Sciences 16,
https://doi.org/10.1007/978-3-031-84977-0_6

particular form for that solution; we just assume that velocities can be expressed as functions of momenta $\dot{\mathbf{q}} = \dot{\mathbf{q}}(\mathbf{q}, \mathbf{p}, t)$ can be found. This statement is actually true for any practical problem since any physically relevant Lagrangian is quadratic in velocities. Let us consider the following function:

$$H(\mathbf{q}, \mathbf{p}, t) = \mathbf{p} \cdot \dot{\mathbf{q}}(\mathbf{q}, \mathbf{p}, t) - L(\mathbf{q}, \dot{\mathbf{q}}(\mathbf{q}, \mathbf{p}, t), t) \,. \tag{6.4}$$

This function is a lot like the energy we defined in Eq. (2.13), except the velocities are replaced by the corresponding functions of momenta. This function is called the *Hamiltonian*. The transformation from the Lagrangian to Hamiltonian defined by (6.4) is one of the examples of the *Legendre transform*, as is explained below. We compute the derivatives of H with respect to $\mathbf{p}$ and $\mathbf{q}$ as

$$\frac{\partial H}{\partial \mathbf{p}} = \dot{\mathbf{q}}(\mathbf{q}, \mathbf{p}, t) + p_\alpha \frac{\partial \dot{q}^\alpha(\mathbf{q}, \mathbf{p}, t)}{\partial \mathbf{p}} - \underbrace{\frac{\partial L}{\partial \dot{q}^\alpha}}_{=p^\alpha} \frac{\partial \dot{q}^\alpha(\mathbf{q}, \mathbf{p}, t)}{\partial \mathbf{p}} = \dot{\mathbf{q}}(\mathbf{q}, \mathbf{p}, t),$$

$$\frac{\partial H}{\partial \mathbf{q}} = p_\alpha \frac{\partial \dot{q}^\alpha(\mathbf{q}, \mathbf{p}, t)}{\partial \mathbf{q}} - \frac{\partial L}{\partial \mathbf{q}} - \underbrace{\frac{\partial L}{\partial \dot{q}^\alpha}}_{=p_\alpha} \frac{\partial \dot{q}^\alpha(\mathbf{q}, \mathbf{p}, t)}{\partial \mathbf{q}} = -\frac{\partial L}{\partial \mathbf{q}} = -\frac{d}{dt}\frac{\partial L}{\partial \dot{\mathbf{q}}} = -\dot{\mathbf{p}}.$$

$$\tag{6.5}$$

Therefore, we have derived Hamilton's equations: given the function $H(\mathbf{q}, \mathbf{p}, t)$ defined by (6.4) for a given Lagrangian $L(\mathbf{q}, \dot{\mathbf{q}}, t)$, the equations of motion for the variables $(\mathbf{q}, \mathbf{p})$ are given by

$$\boxed{\dot{\mathbf{p}} = -\frac{\partial H}{\partial \mathbf{q}}, \qquad \dot{\mathbf{q}} = \frac{\partial H}{\partial \mathbf{p}} \,.} \tag{6.6}$$

Variables $(\mathbf{p}, \mathbf{q})$ are called canonical variables, and Eq. (6.6) is called the canonical equation.

On the Geometric Interpretation of Legendre Transform Let us consider a function $L(v)$ of a single variable $v \in \mathbb{R}$, and take $p = \frac{\partial L}{\partial v}$, as illustrated in Fig. 6.1. At a given point v_0, the variable $p_0 = \frac{\partial L}{\partial v}$ computes the slope of the tangent to the function $L(v)$. The equation of the tangent line is $y = p_0(v - v_0) + L(v_0)$. The intercept of this line with the y-axis, i.e., $v = 0$, is at $y_0 = -p_0 v_0 + L(v_0)$. Thus, the function (Eq. (6.4)) can be understood as the negative of the intercept of the tangent point to the curve with the vertical line. That line is parameterized by the slope p_0 of the tangent instead of the point of contact v_0.

If the function $L(v)$ is convex, the mapping from $L(v)$ to $H(p)$ also produces a convex function. That mapping is called the *Legendre transform*. The inverse Legendre transform is defined by $L(v) = H(p(v)) - p(v)v$, wit $\dot{q} = v = \frac{\partial H}{\partial p}$.

Fig. 6.1 Illustration of the geometric meaning of the Legendre transform as the intercept of the tangent line at $(v_0, L(v_0))$ with the y-axis

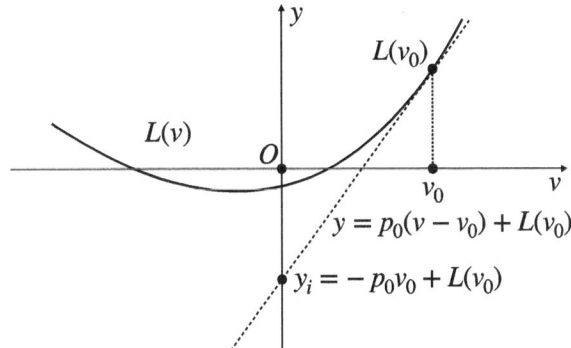

6.2   Examples of Dynamics in Hamiltonian Representation

Here, we will follow the four examples of Sect. 3.7. Please refer to that section for the definition of the problems and notations.

Example 6.1 (A 2D Pendulum) The variable describing the pendulum is φ, the angle with respect to the vertical. For that case, the Lagrangian is given by Eq. (3.18). The momenta and velocities are given by

$$p = \frac{\partial L}{\partial \dot{\varphi}} = ml^2 \dot{\varphi}, \quad \dot{\varphi} = \frac{p}{ml^2}$$

The Hamiltonian is then

$$H = \frac{p^2}{2ml^2} - mgl \cos \varphi.$$

Hamilton's equations are

$$\dot{\varphi} = \frac{\partial H}{\partial p} = \frac{p}{ml^2}, \quad \dot{p} = -\frac{\partial H}{\partial \varphi} = -mgl \sin \varphi.$$

Example 6.2 (2D Elastic Pendulum) The variables describing the motion are l, the length of the pendulum, and φ, the angle of the pendulum with respect to the vertical. As described in Sect. 3.7, the Lagrangian is

$$L = \frac{m}{2} \left(\dot{l}^2 + l^2 \dot{\varphi}^2 \right) + mgl \cos \varphi - U(l).$$

Then, the momenta and velocities expressed in terms of momenta are

$$
\begin{cases}
p_l = \dfrac{\partial L}{\partial \dot{l}} = m\dot{l}, \\[2mm]
p_\varphi = ml^2 \dot{\varphi}.
\end{cases}
\quad \Rightarrow \quad
\begin{cases}
\dot{l} = \dfrac{p_l}{m}, \\[2mm]
\dot{\varphi} = \dfrac{p_\varphi}{ml^2}.
\end{cases}
$$

The Hamiltonian is then

$$
H = \frac{p_l^2}{2m} + \frac{p_\varphi^2}{2ml^2} - mgl \cos \varphi + U(l),
$$

and Hamilton's equations are

$$
\dot{l} = \frac{\partial H}{\partial p_l} = \frac{p_l}{m},
$$

$$
\dot{\varphi} = \frac{\partial H}{\partial p_\varphi} = \frac{p_\varphi}{ml^2}
$$

$$
\dot{p}_l = -\frac{\partial H}{\partial l} = \frac{p_\varphi^2}{ml^3} + mg \cos \varphi - U'(l).
$$

$$
\dot{p}_\varphi = -\frac{\partial H}{\partial \varphi} = -mgl \sin \varphi.
$$

Example 6.3 (Bead on a Hoop of Radius R Rotating with a Constant Angular Velocity ω About the Vertical Axis) The dynamics is described by the position of the bead with respect to the vertical φ. The Lagrangian is derived in Sect. 3.7 as

$$
L = \frac{1}{2} mR^2 \left(\omega^2 \sin^2 \varphi + \dot{\varphi}^2 \right) + mgR \cos \varphi.
$$

The momentum is linearly related to the velocity, and the momentum-velocity relationship is easy to invert:

$$
p = \frac{\partial L}{\partial \dot{\varphi}} = mR^2 \dot{\varphi} \quad \Rightarrow \quad \dot{\varphi} = \frac{p}{mR^2}.
$$

The Hamiltonian is then given by

$$
H = \frac{p^2}{2mR^2} - \frac{1}{2} mR^2 \omega^2 \sin^2 \varphi - mgR \cos \varphi
$$

Then, Hamilton's equations are

$$
\dot{\varphi} = \frac{\partial H}{\partial p} = \frac{p}{mR^2}, \quad \dot{p} = -\frac{\partial H}{\partial \varphi} = -mR^2 \sin \varphi \left(\frac{g}{R} - \omega^2 \cos \varphi \right)
$$

Example 6.4 (A Bead on a Freely Spinning Hoop) The system is described by two variables, the angle of the bead with respect to the vertical φ and the rotation angle of the hoop θ. As derived in Sect. 3.7, the Lagrangian is

$$L = \frac{1}{2} m R^2 (\dot{\theta}^2 \sin^2 \varphi + \dot{\varphi}^2) + m g R \cos \varphi .$$

For $\varphi \neq 0$, the momenta and velocities are given by

$$\begin{cases} p_\theta = \frac{\partial L}{\partial \dot{\theta}} = m R^2 \dot{\theta} \sin^2 \varphi , \\ p_\varphi = m R^2 \dot{\varphi} . \end{cases} \Rightarrow \begin{cases} \dot{\theta} = \frac{p_\theta}{m R^2 \sin^2 \varphi} , \\ \dot{\varphi} = \frac{p_\varphi}{m R^2} . \end{cases}$$

The Hamiltonian is

$$H = \frac{p_\theta^2}{2m R^2 \sin^2 \varphi} + \frac{p_\varphi^2}{2m R^2} - m g R \cos \varphi.$$

Hamilton's equations are

$$\dot{\theta} = \frac{\partial H}{\partial p_\theta} = \frac{p_\theta}{m R^2 \sin^2 \theta},$$

$$\dot{\varphi} = \frac{\partial H}{\partial p_\theta} = \frac{p_\varphi}{m R^2},$$

$$\dot{p}_\theta = -\frac{\partial H}{\partial \theta} = 0,$$

$$\dot{p}_\varphi = -\frac{\partial H}{\partial \varphi} = \frac{p_\theta^2 \cos \varphi}{\sin^3 \varphi} - m g l \sin \varphi.$$

As expected, p_θ is conserved since θ is a cyclic variable.

On Computation of the Hamiltonian for General Quadratic Lagrangians
While the transition from the Lagrangian to the Hamiltonian description is defined for an arbitrary Lagrangian $L(\mathbf{q}, \dot{\mathbf{q}}, t)$ that is convex in velocities $\dot{\mathbf{q}}$, most of the Lagrangians you will encounter in mechanics will be quadratic in velocities, i.e., will have the form

$$L = \frac{1}{2} \dot{\mathbf{q}}^T \mathbb{M}(\mathbf{q}) \dot{\mathbf{q}} - U(\mathbf{q}) = \frac{1}{2} \dot{\mathbf{q}} \cdot \mathbb{M}(\mathbf{q}) \dot{\mathbf{q}} - U(\mathbf{q}) . \tag{6.7}$$

Here, $U(\mathbf{q})$ is some effective potential energy, and we defined a scalar product for two vectors $\mathbf{a}$ and $\mathbf{b}$ in $\mathbb{R}^n$ in the usual way:

$$\mathbf{a} \cdot \mathbf{b} = a_1 b_1 + \ldots a_n b_n = \mathbf{a}^T \mathbf{b} .$$

The matrix $\mathbb{M}(\mathbf{q})$ in Eq. (6.7) is a nonsingular, symmetric "mass matrix" such that $\mathbb{M}^{-1}(\mathbf{q})$ exists, except, perhaps, at some isolated values of $\mathbf{q}$. In the examples above, $\mathbb{M}(\mathbf{q})$ was a diagonal matrix that was easy to invert. In general, for more complex kinematics, $\mathbb{M}(\mathbf{q})$ will need to be explicitly inverted to find the relationship between the momenta and velocities. For quadratic Hamiltonians, we can find a general formula for transferring to the Hamiltonian representation. The momenta and velocity-to-momenta maps are computed as

$$\mathbf{p} = \frac{\partial L}{\partial \dot{\mathbf{q}}} = \mathbb{M}(\mathbf{q})\dot{\mathbf{q}}, \quad \dot{\mathbf{q}} = \mathbb{M}^{-1}(\mathbf{q})\mathbf{p}. \tag{6.8}$$

Then, using the fact that $\mathbb{M}(\mathbf{q}) = \mathbb{M}^T(\mathbf{q})$, we also have $\left(\mathbb{M}^{-1}(\mathbf{q})\right)^T = \mathbb{M}^{-1}(\mathbf{q})$. The Hamiltonian is then computed as

$$H(\mathbf{p}, \mathbf{q}) = \mathbf{p}^T \mathbb{M}^{-1}(\mathbf{q})\mathbf{p} - \frac{1}{2}\mathbf{p}^T \left(\mathbb{M}^{-1}(\mathbf{q})\right)^T \mathbb{M}(\mathbf{q})\mathbb{M}^{-1}\mathbf{p} + U(\mathbf{q})$$
$$= \frac{1}{2}\mathbf{p}^T \mathbb{M}^{-1}(\mathbf{q})\mathbf{p} + U(\mathbf{q}). \tag{6.9}$$

Thus, in the case when the Lagrangian is given as a quadratic function of the velocity with an arbitrary function of $U(\mathbf{q})$, we only need to compute the inverse of the mass matrix $\mathbb{M}^{-1}(\mathbf{q})$ to determine the Hamiltonian (Eq. (6.9)). If we denote $\mathbb{B}(\mathbf{q}) = \mathbb{M}^{-1}(\mathbf{q})$, then Hamilton's equations are

$$\dot{\mathbf{q}} = \frac{\partial H}{\partial \mathbf{p}} = \mathbb{B}(\mathbf{q})\mathbf{p}, \quad \dot{\mathbf{p}} = -\frac{\partial H}{\partial \mathbf{q}} = -\frac{1}{2}\mathbf{p}^T \frac{\partial \mathbb{B}(\mathbf{q})}{\partial \mathbf{q}}\mathbf{p} - \frac{\partial U(\mathbf{q})}{\partial \mathbf{q}}$$

or, in coordinates,

$$\dot{q}^i = B^{ij}p_j, \quad \dot{p}_i = -\frac{1}{2}p_j \frac{\partial B^{jk}}{\partial q^i}p_k - \frac{\partial U}{\partial q^i}.$$

6.3 Cotangent Bundle

We have talked about the tangent bundle in connection with the Lagrangian description of motion. In the Hamiltonian description, the variables are $\mathbf{q} \in Q$ and $\mathbf{p}$. Unlike the $\dot{\mathbf{q}}$, which have indices up, the variables $\mathbf{p}$ have indices down, so the momenta are covectors. The mathematical construction allowing the formalization of this approach is called *the cotangent bundle* of a manifold.

A cotangent space at a point q of Q, denoted as T_q^*Q, is the set of all covectors dual to the vector space $T_q Q$. The union of all cotangent spaces $\cup_{q \in Q} T_q^*Q$ is called the cotangent bundle, T^*Q. Mathematically, $(\mathbf{q}, \mathbf{p})$ are local coordinates on the cotangent bundle.

If we introduce the coordinate $\boldsymbol{\xi} = (\mathbf{q}, \mathbf{p})$, so $\xi_i = q_i$, $(i = 1, \ldots, n)$ and $\xi_i = p_{i-n}$, $(i = n + 1, \ldots 2n)$, then Hamilton's equations are written as

$$\dot{\xi}^i = \omega^{ij} \frac{\partial H}{\partial \xi^i}, \quad \Omega = \left\{ \omega^{ij} \right\}_{i,j=1\ldots n}, \quad \Omega = \begin{pmatrix} 0_{n \times n} & \mathrm{Id}_{n \times n} \\ -\mathrm{Id}_{n \times n} & 0_{n \times n} \end{pmatrix}, \qquad (6.10)$$

so Ω is a matrix composed of entries ω^{ij}. Note that for simplicity, we wrote all indices of ξ^i to be up even though some of ξ^i is composed of the momenta variables. The matrix Ω is called the symplectic matrix. It has the following properties:

$$\Omega^T = -\Omega, \quad \Omega^2 = \Omega\Omega = -\mathrm{Id}_{2n \times 2n}. \qquad (6.11)$$

6.4 Poisson Brackets

Suppose we are interested in the rate of change of some function $F(\mathbf{q}, \mathbf{p}, t)$, and $\mathbf{q}(t)$ and $\mathbf{p}(t)$ satisfy the canonical equations (Eq. (6.6)) with the Hamiltonian $H(\mathbf{q}, \mathbf{p}, t)$. We have

$$\frac{dF}{dt} = \frac{\partial F}{\partial \mathbf{q}} \cdot \dot{\mathbf{q}} + \frac{\partial F}{\partial \mathbf{p}} \cdot \dot{\mathbf{p}} + \frac{\partial F}{\partial t} = \frac{\partial F}{\partial \mathbf{q}} \cdot \frac{\partial H}{\partial \mathbf{p}} - \frac{\partial F}{\partial \mathbf{p}} \cdot \frac{\partial H}{\partial \mathbf{q}} + \frac{\partial F}{\partial t}. \qquad (6.12)$$

Let us introduce the new operator $\{,\}$ that takes two functions of $F(\mathbf{q}, \mathbf{p})$ and $G(\mathbf{q}, \mathbf{p})$ (or possibly $F(\mathbf{q}, \mathbf{p}, t)$ and $G(\mathbf{q}, \mathbf{p}, t)$), producing another function of $(\mathbf{q}, \mathbf{p})$ and possibly t. This operator is called the canonical Poisson bracket:

$$\{F, G\} = \frac{\partial F}{\partial \mathbf{q}} \cdot \frac{\partial G}{\partial \mathbf{p}} - \frac{\partial F}{\partial \mathbf{p}} \cdot \frac{\partial G}{\partial \mathbf{q}} = \frac{\partial F}{\partial q^\alpha} \frac{\partial G}{\partial p_\alpha} - \frac{\partial F}{\partial p_\alpha} \frac{\partial G}{\partial q^\alpha}. \qquad (6.13)$$

We have $\{F, G\} = -\{G, F\}$, so the bracket is an antisymmetric operator. Moreover, Eq. (6.12) is equivalent to

$$\frac{dF}{dt} = \{F, H\} + \frac{\partial F}{\partial t}. \qquad (6.14)$$

One can immediately see several useful consequences arising from Eq. (6.14):

1. If F does not explicitly depend on time, then $\dot{F} = \{F, H\}$.
2. For any function H, $\{H, H\} = 0$ since the bracket is antisymmetric. Thus, if H does not depend on time, H is a constant of motion.
3. A function F that is time-independent is a constant of motion if and only if $\{F, H\} = 0$.

So, computing constants of motion for a Hamiltonian system seems "easy": all we need to do is to find a time-independent function F such that $\{F, H\} = 0$. There is no need to verify the actual conservation of F! However, unfortunately, finding this function is not that easy, although sometimes one can notice certain patterns for particular Hamiltonians that yield $\{F, H\} = 0$. It is definitely not guaranteed for any H, but if you can notice a conserved quantity (apart from H if it is time-independent), please do it.

General Poisson Systems Next, we turn our attention to the properties of the Poisson brackets. For any (twice) differentiable functions $F(\mathbf{q}, \mathbf{p})$, $G(\mathbf{q}, \mathbf{p})$, and $H(\mathbf{q}, \mathbf{p})$, the following properties are true:

1. Antisymmetry: $\{F, G\} = -\{G, F\}$ (proved by definition).
2. Bi-linearity: $\{\alpha F + \beta G, H\} = \alpha\{F, H\} + \beta\{G, H\}$ and similarly with the second argument.
3. Derivative-like (satisfies Leibnitz rule): $\{FG, H\} = F\{G, H\} + G\{F, H\}$ (proved by definition).
4. Jacobi identity:

$$\{F, \{G, H\}\} + \{H, \{F, G\}\} + \{G, \{H, F\}\} = 0. \tag{6.15}$$

The Jacobi identity (Eq. (6.15)) for the canonical bracket (Eq. (6.13)) is derived by a straightforward—but tedious—differentiation and use of formula (Eq. (6.13)). For a general Poisson bracket, it is a requirement that the potential Poisson bracket has to satisfy the four properties above. The Jacobi identity is actually the hardest to verify for any potential Poisson bracket.

Manifolds endowed with a Poisson bracket are called *Poisson*. If the bracket has the form (Eq. (6.13)), these systems are called *canonical*; otherwise, the system is called noncanonical. Noncanonical Poisson systems are actually quite common; they often arise from canonical systems due to the reduction of symmetry. Many systems describing satellite dynamics, fluids, plasmas, and many other physical systems can be written as the *Lie-Poisson systems*; see [4] (Chap. 10), [3] (Chap. 2), or [48] (Chap. 9).

For general Poisson brackets, the variables in the phase space are no longer separated into coordinates and momenta. Sometimes, physical meaning can be assigned to the coordinates, especially in the Lie-Poisson case. Sometimes, noncanonical brackets have special functions C called *Casimir invariants* (or simply Casimirs) that Poisson-commute with *any function of the phase space*: $\{C, H\} = 0$ *for any* function H. These Casimir invariants are a property of the bracket and, if they exist, play a much deeper role in the dynamics compared to a particular choice of the Hamiltonian.

Alternative Form of Hamilton's Equation Hamilton's equations (Eq. (6.6)) can be written in terms of the canonical bracket:

$$\dot{p}_\alpha = \{p_\alpha, H\}, \quad \dot{q}^\alpha = \{q^\alpha, H\}. \tag{6.16}$$

The easiest way to verify this relationship is to notice that

$$\frac{\partial p_\alpha}{\partial p_\beta} = \delta_\alpha^\beta = \begin{cases} 1, & \alpha = \beta \\ 0, & \alpha \neq \beta \end{cases} \quad \text{and} \quad \frac{\partial q^\alpha}{\partial q^\beta} = \delta_\beta^\alpha = \begin{cases} 1, & \alpha = \beta \\ 0, & \alpha \neq \beta \end{cases} \tag{6.17}$$

Lemma 6.1 (Conserved Quantities) *Suppose a quantity $F(\mathbf{q}, \mathbf{p})$ is such that for a given Hamiltonian, $\{F, H\} = 0$. Then, F is conserved on all the solutions of Hamilton's equations (Eq. (6.16)), i.e., $F(\mathbf{p}(t), \mathbf{q}(t)) = F(\mathbf{p}(0), \mathbf{q}(0))$*

Proof On the solutions of Hamilton's equations, since F does not explicitly depend on time,

$$\frac{dF}{dt} = \{F, H\} = 0 \quad \Rightarrow \quad F = \text{const.} \tag{6.18}$$

Corollary *If the Hamiltonian does not explicitly depend on time, it is conserved on solutions. Indeed, $\{H, H\} = 0$ because the Poisson bracket is antisymmetric.*

Theorem 6.1 (Poisson's Theorem: Creating New Conserved Quantities) *Suppose $F(\mathbf{q}, \mathbf{p}, t)$ and $G(\mathbf{q}, \mathbf{p}, t)$ satisfy $\{F, H\} = 0$ and $\{G, H\} = 0$. Then, the quantity $Q = \{F, G\}$ also satisfies $\{Q, H\} = 0$.*

Proof Using the Jacobi identity for Poisson brackets,

$$\{Q, H\} = \{\{F, G\}, H\} = -\{\{H, G\}, F\} - \{\{F, H\}, G\} = 0. $$

∎

If $F(\mathbf{q}, \mathbf{p})$ and $G(\mathbf{q}, \mathbf{p})$ do not depend explicitly on time, then according to Eq. (6.18), F and G are constants of motion. Then $Q = \{F, G\}$ is also a constant of motion.

Poisson's theorem allows us to find new conserved quantities from the existing conserved quantities. One can then take $\{Q, F\}$ and $\{Q, G\}$ and generate new conserved quantities and so on. Clearly, one cannot keep generating these conserved quantities indefinitely according to Poisson's theorem; otherwise, every Hamiltonian system that has two conserved quantities independent of the Hamiltonian will have an infinite number of conserved quantities. Thus, sometimes $\{F, G\}$ does yield a new conserved quantity, and sometimes it gives $\{F, G\} = 0$; that is where the procedure stops. Still, Poisson's theorem can yield useful new constants of motion and can be extremely beneficial in some cases.

6.5 Examples of Poisson Bracket Computations

Example 6.1 Consider an n-dimensional problem $Q = \mathbb{R}^n$, with $\mathbf{q} = (q_1, \ldots, q_n)$ and $\mathbf{p} = (p_1, \ldots, p_n)$ (we keep all indices down for simplicity). Suppose $\xi = |\mathbf{q}|^2 - |\mathbf{p}|^2$ and $\eta = \mathbf{q} \cdot \mathbf{p}$. Compute the Poisson bracket $\{\xi, \eta\}$:

$$\{\xi, \eta\} = 2\mathbf{q} \cdot \mathbf{q} + 2\mathbf{p} \cdot \mathbf{p} = 2\left(|\mathbf{q}|^2 + |\mathbf{p}|^2\right). \tag{6.19}$$

Example 6.2 Consider the same n-dimensional problem as in Example 6.1, and take $F(\mathbf{q}, \mathbf{p}) = A(\xi, \eta)$ and $G(\mathbf{q}, \mathbf{p}) = B(\xi, \eta)$, where A and B are smooth functions of their arguments. Find the Poisson bracket $\{F, G\}$.

Since the Poisson bracket satisfies the Leibnitz rule, we use the result (Eq. (6.19)) and compute the Poisson bracket as follows:

$$\begin{aligned}
\{F, G\} &= \frac{\partial A}{\partial \xi}\frac{\partial B}{\partial \xi}\{\xi, \xi\} + \frac{\partial A}{\partial \eta}\frac{\partial B}{\partial \xi}\{\eta, \xi\} + \frac{\partial A}{\partial \xi}\frac{\partial B}{\partial \eta}\{\xi, \eta\} + \frac{\partial A}{\partial \eta}\frac{\partial B}{\partial \eta}\{\eta, \eta\} \\
&= \left(\frac{\partial A}{\partial \xi}\frac{\partial B}{\partial \eta} - \frac{\partial A}{\partial \eta}\frac{\partial B}{\partial \xi}\right)\{\xi, \eta\} = 2\left(|\mathbf{q}|^2 + |\mathbf{p}|^2\right)\left(\frac{\partial A}{\partial \xi}\frac{\partial B}{\partial \eta} - \frac{\partial A}{\partial \eta}\frac{\partial B}{\partial \xi}\right),
\end{aligned}$$

where we have used $\{\xi, \xi\} = \{\eta, \eta\} = 0$, and $\{\xi, \eta\} = -\{\eta, \xi\}$.

Example 6.3 (Angular Momentum) Let us compute the Poisson bracket for the following interesting example. Suppose $Q = \mathbb{R}^3$ and the cotangent bundle can be described by two vectors, $\mathbf{q} \in \mathbb{R}^3$ and $\mathbf{p} \in \mathbb{R}^3$. Take the vector of angular momentum $\mathbf{L} = \mathbf{q} \times \mathbf{p}$. Take a fixed vector $\mathbf{v} \in \mathbb{R}^3$, and form $F = \mathbf{L} \cdot \mathbf{v}$. Find $\{F, G\}$ for some arbitrary function $G(\mathbf{q}, \mathbf{p})$. We use the property of mixed product identities to put the differentiated quantities first:

$$\begin{aligned}
\{F, G\} &= \frac{\partial}{\partial q^\alpha}\underbrace{(\mathbf{q} \times \mathbf{p}) \cdot \mathbf{v}}_{\text{rotate}}\frac{\partial G}{\partial p_\alpha} - \frac{\partial}{\partial p_\alpha}(\mathbf{q} \times \mathbf{p}) \cdot \mathbf{v}\frac{\partial G}{\partial q^\alpha} \\
&= \frac{\partial}{\partial q^\alpha}\mathbf{q} \cdot (\mathbf{p} \times \mathbf{v})\frac{\partial G}{\partial p_\alpha} - \frac{\partial}{\partial p_\alpha}\mathbf{p} \cdot (\mathbf{v} \times \mathbf{q})\frac{\partial G}{\partial q^\alpha} \qquad (6.20) \\
&= (\mathbf{p} \times \mathbf{v}) \cdot \frac{\partial G}{\partial \mathbf{p}} - (\mathbf{v} \times \mathbf{q}) \cdot \frac{\partial G}{\partial \mathbf{q}}.
\end{aligned}$$

Now, take $G = \mathbf{L} \cdot \mathbf{w} = (\mathbf{q} \times \mathbf{p}) \cdot w\mathbf{w}$, where $\mathbf{w}$ is another fixed vector. Then, using the same identities for mixed products, we obtain

$$\begin{aligned}
\frac{\partial G}{\partial \mathbf{q}} &= \frac{\partial}{\partial \mathbf{q}}(\mathbf{q} \times \mathbf{p}) \cdot \mathbf{w} = \frac{\partial}{\partial \mathbf{q}}\mathbf{q} \cdot (\mathbf{p} \times \mathbf{w}) = \mathbf{p} \times \mathbf{w}, \\
\frac{\partial G}{\partial \mathbf{p}} &= \frac{\partial}{\partial \mathbf{p}}(\mathbf{q} \times \mathbf{p}) \cdot \mathbf{w} = \frac{\partial}{\partial \mathbf{p}}\mathbf{p} \cdot (\mathbf{w} \times \mathbf{q}) = \mathbf{w} \times \mathbf{q}.
\end{aligned} \tag{6.21}$$

Substitution of Eq. (6.21) into Eq. (6.20), we find

$$
\begin{aligned}
\{\mathbf{L} \cdot \mathbf{v}, \mathbf{L} \cdot \mathbf{w}\} &= (\mathbf{p} \times \mathbf{v}) \cdot (\mathbf{w} \times \mathbf{q}) - (\mathbf{v} \times \mathbf{q}) \cdot (\mathbf{p} \times \mathbf{w}) \\
&= \mathbf{w} \cdot (\ \underbrace{\mathbf{q} \times (\mathbf{p} \times \mathbf{v}) + \mathbf{p} \times (\mathbf{v} \times \mathbf{q})}_{\text{the Jacobi identity for} \times \text{product}}\ ) \\
&= -\mathbf{w} \cdot (\mathbf{v} \times (\mathbf{q} \times \mathbf{p})) = \mathbf{L} \cdot (\mathbf{v} \times \mathbf{w}).
\end{aligned}
\tag{6.22}
$$

I invite you to appreciate the mathematical elegance of this result. In particular, if we take $\mathbf{v} = \mathbf{e}_\alpha$ and $\mathbf{w} = \mathbf{e}_\beta$, the basis vectors, then $\mathbf{v} \times \mathbf{w} = \epsilon_{\alpha\beta\gamma} \mathbf{e}_\gamma$, so the Poisson bracket of components of angular momenta is given by $\{L_\alpha, L_\beta\} = \epsilon_{\alpha\beta\gamma} L_\gamma$.[1]

6.6 Looking Forward

The Hamiltonian approach presents an approach to mechanics just as powerful as the Lagrangian approach. The exposition of Hamiltonian mechanics is often considered from the point of view of geometry, as it includes differential forms, vector fields, and manifolds. We will touch upon these concepts in the next section. There are many excellent introductions to Hamiltonian mechanics [1–4, 24, 49, 50]. In the next several chapters, we will build the mathematical structures underpinning Hamiltonian mechanics and show how these mathematical structures are useful in designing new concepts such as canonical transformations and Hamilton-Jacobi equations.

One can pose a question of whether the Lagrangian or Hamiltonian approach to mechanics is better. There is no unique answer to that question, as it depends on the problem considered. Some questions are more easily treated in the Lagrangian framework and some in the Hamiltonian framework. You should aim to be equally proficient in both approaches to be ready to solve a wide variety of problems that you may encounter in the future.

6.7 Practice Problems

1. Find the Hamiltonian for the following Lagrangians with $\mathbf{q} \in \mathbb{R}^n$ and write Hamilton's equations:

---

[1] We used the notation $\epsilon_{\alpha\beta\gamma}$ to denote the completely antisymmetric Levi-Civita symbol, with the convention that if any of the indices α, β, γ are equal, $\epsilon_{\alpha\beta\gamma} = 0$. If none of the indices are equal, then $\epsilon_{123} = 1$, and any it also equals to 1 for any even permutation of indices $(1, 2, 3)$, and -1 or any odd permutation of indices $(1, 2, 3)$.

(a)

$$L(\dot{\mathbf{q}}, \mathbf{q}) = \frac{|\dot{\mathbf{q}}|^2}{2|\mathbf{q}|^2} - U(\mathbf{q}).$$

(b)

$$L(\dot{\mathbf{q}}, \mathbf{q}) = \frac{1}{2}\dot{\mathbf{q}}^T \mathbb{A}(\mathbf{q})\dot{\mathbf{q}} + \mathbf{F}(\mathbf{q}) \cdot \dot{\mathbf{q}}.$$

$\mathbb{A}(\mathbf{q})$ is a non-singular, symmetric $n \times n$ matrix, and $\mathbf{F}(\mathbf{q})$ is a vector-valued function with the values in $\mathbb{R}^n$. Assume that the inverse of $\mathbb{A}(\mathbf{q})$ is known; write your answer using $\mathbb{B} = \mathbb{A}^{-1}(\mathbf{q})$.

Note. In writing Hamilton's equations, you will need to compute the derivative of the inverse matrix with respect to coordinates $\frac{\partial \mathbb{B}(\mathbf{q})}{\partial(\mathbf{q})}$. It is OK to just keep this expression implicitly as is; assuming you can compute $\mathbb{B}(\mathbf{q}) = \mathbb{A}^{-1}(\mathbf{q})$, you can also compute the derivatives of that matrix. If you want to express the answer explicitly in terms of $\mathbb{A}$, you can use the following formula:

$$\frac{\partial \mathbb{A}^{-1}(\mathbf{q})}{\partial q^i} = -\mathbb{A}^{-1}(\mathbf{q})\frac{\partial \mathbb{A}(\mathbf{q})}{\partial q^i}\mathbb{A}^{-1}(\mathbf{q}).$$

Proving this formula is a useful exercise by itself.

2. For this problem, do the following:

   (a) Identify the configuration manifold and compute all the relevant coordinates and velocities.
   (b) Compute the Lagrangian.
   (c) Using the Legendre transform, compute the Hamiltonian.
   (d) Write Hamilton's equations.

   (a) **Pendulum on a cart.** Consider a cart that can move freely along the horizontal axis. The cart is carrying a pendulum of the length l, and the mass of the bob is m; the mass of the cart is M. This problem is illustrated on the left panel of Fig. 6.2.
   (b) **Two masses on intersecting wires.** Consider two masses m_1 and m_2, connected with a massless rigid rod of length l. The masses move freely along the guiding wires; one mass can only move along the x-axis, and the second mass can only move along the y-axis. Each of the masses can freely pass through the intersection point of the wires without changing direction or encountering additional friction. Neglect gravity. This problem is illustrated on the right panel of Fig. 6.2.

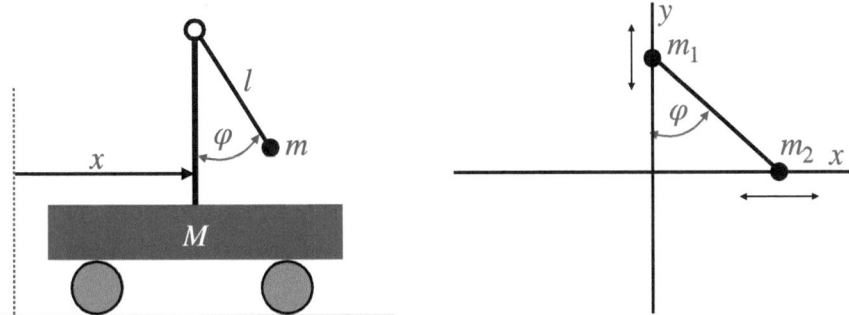

Fig. 6.2 Left: a pendulum on a freely moving cart. Right: two masses that can move on perpendicular axes, connected with a rigid rod of length l. The first mass can only move along the x-axis, and the second mass can move only along the y-axis

3. Compute the canonical Poisson bracket of the following functions:

   (a) One-dimensional system, $F = p^2 + q^2$ and $G = \arctan(p/q)$
   (b) Two-dimensional system, $F = p_1^2 + q_2^2$ and $G = p_2^2 + q_1^2 + \varphi(q_1^2 + q_2^2)$, where $\varphi(x)$ is an arbitrary function of the scalar argument x.
   (c) N-dimensional system: Suppose $\varphi(x)$ is an arbitrary function of a scalar variable x and $F(\mathbf{p}, \mathbf{q})$ is an arbitrary function of $(\mathbf{p}, \mathbf{q})$. Compute the Poisson bracket of $F(\mathbf{p}, \mathbf{q})$ and $G(\mathbf{p}, \mathbf{q}) = \varphi(F(\mathbf{p}, \mathbf{q}))$.

4. Conservation laws:

   (a) Suppose a canonical system has the Hamiltonian that is written in the form

$$H(\mathbf{p}, \mathbf{q}) = F(\varphi_1(p_1, q_1), \varphi_2(p_2, q_2) \ldots, \varphi_n(p_n, q_n)),$$

   where $F(x_1, x_2, \ldots, x_n)$ is an arbitrary function of n variables, and $\varphi_i(p_i, q_i)$ are arbitrary functions of two variables. Show that $\{\varphi_i, H\} = 0$, so each of $\varphi_i(p_i, q_i)$ is a constant of motion.

   (b) Suppose a canonical system has a Hamiltonian that can be written as

$$H(\mathbf{p}, \mathbf{q}) = \frac{\sum_i f_i(p_i, q_i)}{\sum_i \varphi_i(p_i, q_i)}.$$

   Show that the system has n first integrals of the form $P_i = f_i(p_i, q_i) - H\varphi_i(p_i, q_i)$.

Chapter 7
Introduction to Differential Forms and Exterior Calculus

7.1 Vector Fields, One-Forms, and Exterior Products

Before we proceed any further, we need to introduce some modern language of vector fields and differential forms. This language is not an abstraction for abstraction's sake: using this language, the discussions of canonical transformations, integral invariants, Hamilton-Jacobi equations, and other topics become elegant and simple to understand. That is why most modern studies in mechanics use this mathematical language. We will learn the basics of this language without much abstraction. If you want to continue your studies on this subject, you will need to further extend this knowledge to include the theory of Lie groups and manifolds [4].

Vector Fields We introduce the concept of vector fields through ODEs. Suppose $\boldsymbol{\xi}$ are the local coordinates on some smooth manifold M, e.g., describing the cotangent bundle of the configuration manifold, so $\boldsymbol{\xi} = (\mathbf{q}, \mathbf{p})$. Suppose there is an ODE given by $\dot{\boldsymbol{\xi}} = \mathbf{f}(\boldsymbol{\xi})$. Take an arbitrary function $g(\boldsymbol{\xi})$. What is the rate of change of $g(\boldsymbol{\xi})$ on solutions of our ODE? We know that (remember always to sum over repeated indices!)

$$\frac{dg}{dt} = \frac{\partial g}{\partial \xi^\alpha}\dot{\xi}^\alpha = f^\alpha \frac{\partial g}{\partial \xi^\alpha} = \left(f^\alpha \frac{\partial}{\partial \xi^\alpha}\right)g\,. \tag{7.1}$$

Let us isolate the object in parentheses above acting on g. We call it a *vector field* on M. Vector fields are usually denoted by capital letters such as X, Y, and Z and can be written in local coordinates on M as

$$\text{Vector field } X: \quad X^\alpha(\boldsymbol{\xi})\frac{\partial}{\partial \xi^\alpha}\,. \tag{7.2}$$

© The Author(s), under exclusive license to Springer Nature Switzerland AG 2025
V. Putkaradze, *A Concise Introduction to Classical Mechanics*,
Surveys and Tutorials in the Applied Mathematical Sciences 16,
https://doi.org/10.1007/978-3-031-84977-0_7

Vector fields are defined on the TM, the tangent bundle to M, since for every $\boldsymbol{\xi}$, every vector field (7.2) defines a tangent vector in $T_{\boldsymbol{\xi}} M$. Clearly, the space of all vector fields on a manifold is a linear space. Vector fields act on functions on M by taking gradients in the prescribed direction:

$$X(g) = X^\alpha \frac{\partial g}{\partial \xi^\alpha} . \tag{7.3}$$

Commutator of Vector Fields Let us see what happens when we apply two vector fields, X and Y, to a function f. In local coordinates, we get

$$
\begin{aligned}
X &= X^\alpha(\boldsymbol{\xi}) \frac{\partial}{\partial \xi^\alpha}, \quad Y = Y^\alpha(\boldsymbol{\xi}) \frac{\partial}{\partial \xi^\alpha}, \\
X(Y(f)) &= X^\beta \frac{\partial Y^\alpha}{\partial \xi^\beta} \frac{\partial f}{\partial \xi^\alpha} + X^\beta Y^\alpha \frac{\partial^2 f}{\partial \xi^\alpha \partial \xi^\beta}, \\
Y(X(f)) &= Y^\beta \frac{\partial X^\alpha}{\partial \xi^\beta} \frac{\partial f}{\partial \xi^\alpha} + Y^\beta X^\alpha \frac{\partial^2}{f} \partial \xi^\alpha \partial \xi^\beta, \\
X(Y(f)) - Y(X(f)) &= X^\beta \frac{\partial Y^\alpha}{\partial \xi^\beta} \frac{\partial f}{\partial \xi^\alpha} - Y^\beta \frac{\partial X^\alpha}{\partial \xi^\beta} \frac{\partial f}{\partial \xi^\alpha} .
\end{aligned}
\tag{7.4}
$$

Thus, neither $X(Y)$ nor $Y(X)$ are a vector field since their action on a function involves second derivatives. However, the combination $X(Y(f)) - Y(X(f))$ is a vector field since all second derivatives cancel. That vector field is known as the *commutator* $[X, Y]$ and, according to Eq. (7.4), is defined as follows:

$$[X, Y](f) = X^\beta \frac{\partial Y^\alpha}{\partial \xi^\beta} \frac{\partial f}{\partial \xi^\alpha} - Y^\beta \frac{\partial X^\alpha}{\partial \xi^\beta} \frac{\partial f}{\partial \xi^\alpha} . \tag{7.5}$$

We can prove by direct calculation that the commutator of any three vector fields (X, Y, Z) on a manifold M satisfies the Jacobi identity:

$$[X, [Y, Z]] + [Z, [X, Y]] + [Y, [Z, X]] = 0 . \tag{7.6}$$

This result can be proven by a direct but somewhat tedious computation, which is left to the reader as an exercise. A vector space with an antisymmetric, bilinear commutator operator $[\cdot, \cdot]$, satisfying the Jacobi identity (Eq. (7.6)) is called *a Lie algebra*. Thus, vector fields on a smooth manifold form a Lie algebra.

Other Lie Algebras You have met other Lie algebras before; the most familiar example is $\mathbb{R}^3$. For any $\mathbf{u}, \mathbf{v} \in \mathbb{R}^3$, we define the commutator as the vector product $[\mathbf{u}, \mathbf{v}] = \mathbf{u} \times \mathbf{v}$.

Exterior One-Forms We have seen that it is natural to define vector fields as a result of action on functions. Let us go one step further: what objects act on vector fields? As it turns out, these objects are called one-forms. They are defined as

$$1\text{-form } \theta = \theta_\alpha(\boldsymbol{\xi})d\xi^\alpha \quad (\text{sum over } \alpha). \tag{7.7}$$

It is then natural to define how this one-form acts on any vector field. By definition,

$$\theta(X) := \theta_\alpha(\boldsymbol{\xi})d\xi^\alpha \left(X^\alpha(\boldsymbol{\xi})\frac{\partial}{\partial\xi^\alpha} \right) := \theta_\alpha(\boldsymbol{\xi})X^\alpha(\boldsymbol{\xi}). \tag{7.8}$$

Note that one-forms form a linear space.

Differential Forms and Hamiltonian Vector Fields For vector fields defined on T^*Q, we will separate the vector fields acting on the coordinates q^α and on the momenta p^α and write $X = f^\alpha\frac{\partial}{\partial q^\alpha} + g_\alpha\frac{\partial}{\partial p_\alpha}$. In particular, it is interesting to consider the Hamiltonian vector fields originating from the Hamiltonian H. Since the components of the vector field $\dot{\mathbf{q}}$ and $\dot{\mathbf{p}}$ satisfy the canonical equations (Eq. (6.6)), we define

$$X_H = \frac{\partial H}{\partial p_\alpha}\frac{\partial}{\partial q^\alpha} - \frac{\partial H}{\partial q^\alpha}\frac{\partial}{\partial p_\alpha} \quad (\text{sum over } \alpha). \tag{7.9}$$

For any function $F(\mathbf{q}, \mathbf{p})$,

$$X_H(F) = \{F, H\}. \tag{7.10}$$

Let us now define two one-forms important for mechanics. The first form is defined on variables T^*Q and only has components on the configuration manifold Q:

$$\theta_0 = p_\alpha dq^\alpha \quad (\text{sum over } \alpha). \tag{7.11}$$

As an exercise, compute $\theta_0(X_H)$ for a given $H(\mathbf{q}, \mathbf{p})$. (The answer is $p_\alpha\frac{\partial H}{\partial p_\alpha}$).

Two-Forms Is it possible to create an object that acts on two vector fields? The answer is yes, and it is called a two-form. If M is a smooth manifold, and X and Y are two vector fields, a two-form ω is a bilinear, antisymmetric operator on vector fields (i.e., it produces a scalar function given two vector fields). Thus, a two-form must satisfy

1. $\omega(X, Y) = -\omega(Y, X)$ for any vector fields X and Y.
2. $\omega(aX_1 + bX_2, Y) = a\omega(X_1, Y) + b\omega(X_2, Y)$, for any vector fields X_1, X_2 and Y.

Two-forms also form a linear space. We can write some examples of two-forms:

1. For any two vector fields $X_\alpha\frac{\partial}{\partial\xi^\alpha}$ and $Y_\beta\frac{\partial}{\partial\xi^\beta}$, we define $\omega(X, Y) = \sum_{\alpha,\beta} X_\alpha Y_\beta - X_\beta Y_\alpha$.

2. Suppose $M = \mathbb{R}^3$. Then, vector fields $\mathbf{X}$ and $\mathbf{Y}$ can be identified with three-dimensional vector-valued functions $\mathbf{X}$ and $\mathbf{Y}$. Then, take a fixed vector-valued function $\mathbf{V}$, and define

$$\omega(\mathbf{X}, \mathbf{Y}) = \mathbf{V} \cdot (\mathbf{X} \times \mathbf{Y}). \tag{7.12}$$

When $\mathbf{X}$, $\mathbf{Y}$ and $\mathbf{V}$ are vectors, Eq. (7.12) defines the flux of fluid material carried by vector $\mathbf{V}$ through the surface defined by the parallelogram formed by vectors $(\mathbf{X}, \mathbf{Y})$; see [1], Chap. 39B.

Substitution Operator Suppose we have a two-form, and we want to define the operator that uses only one vector field to "pair" with the two-form. We would expect that the result would need another vector field to make a scalar function. It would be nice if the result of such a "pairing," "substitution," or "inner product" would be a one-form. We accomplish this task by defining the substitution operator i_X. First, for one-form θ and a vector field X, we define simply $i_X\theta = \theta(X)$. Using the local coordinates ξ^α on the manifold M, we write

$$\theta = \theta_\alpha d\xi^\alpha, \quad X = X^\alpha \frac{\partial}{\partial \xi^\alpha}, \quad \Rightarrow \quad i_X\theta = \theta(X) = \theta_\alpha X^\alpha. \tag{7.13}$$

For any two-form ω, given a vector field X, we define a one-form $i_X\omega$ as

$$(i_X\omega)(Y) = \omega(X, Y), \quad \text{for any vector field } Y. \tag{7.14}$$

Exterior Product Let us now compute a way to produce two-forms out of one-forms. If θ and ψ are one-forms, then we define the exterior product of one-forms as a two-form $\theta \wedge \psi$. We can define this form by requiring that the result of the substitution operator on $\theta \wedge \psi$ is a one-form:

$$i_X(\theta \wedge \psi) = (i_X\theta)\psi - (i_X\psi)\theta = \theta(X)\psi - \psi(X)\theta. \tag{7.15}$$

Alternatively, we can compute the action of this form on two vector fields X, Y. Using Eq. (7.15), we get

$$(\theta \wedge \psi)(X, Y) = i_Y\big(i_X(\theta \wedge \psi)\big) = (i_X(\theta \wedge \psi))(Y) = \theta(X)\psi(Y) - \psi(X)\theta(Y). \tag{7.16}$$

In local coordinates, all two-forms look like

$$\omega = \omega_{\alpha\beta}(\xi)d\xi^\alpha \wedge d\xi^\beta, \quad \omega_{\alpha\beta}(\xi) = -\omega_{\beta\alpha}(\xi). \tag{7.17}$$

Exterior Derivatives Let us now define the concept of exterior derivative. For a function $f(\xi)$ (a 0-form), we define a one-form df such that $df(X) = X(f)$, for any vector field X. That may look a little abstract, but writing this derivative in local coordinates makes perfect sense since

$$\mathrm{d}f(X) = \frac{\partial f}{\partial \xi^\alpha} \mathrm{d}\xi^\alpha .$$ (7.18)

In order to define the (exterior) derivative of one-form defined as $\theta = \theta_\alpha(\boldsymbol{\xi})\mathrm{d}\xi^\alpha$, we posit

$$\mathrm{d}\theta = \mathrm{d}\left(\theta_\alpha(\boldsymbol{\xi})\mathrm{d}\xi^\alpha\right) = \mathrm{d}\theta_\alpha(\boldsymbol{\xi}) \wedge \mathrm{d}\xi^\alpha = \frac{\partial \theta_\alpha}{\partial \xi^\beta} \mathrm{d}\xi^\beta \wedge \mathrm{d}\xi^\alpha .$$ (7.19)

In particular, the form θ_0 that we have defined in Eq. (7.7) has the following exterior derivative:

$$\mathrm{d}\theta_0 = \mathrm{d}(p_\alpha \mathrm{d}q^\alpha) = \mathrm{d}p_\alpha \wedge \mathrm{d}q^\alpha .$$ (7.20)

This two-form will play an important role and (with the change of the sign) is called the canonical two-form:

$$\omega = \mathrm{d}q^\alpha \wedge \mathrm{d}p_\alpha = -\mathrm{d}\theta_0 .$$ (7.21)

Closed and Exact Forms   Let us also compute the two-form $\mathrm{d}(\mathrm{d}f)$ for any function $f(\boldsymbol{\xi})$. We have

$$\begin{aligned}
\mathrm{d}(\mathrm{d}f) &= \mathrm{d}\left(\frac{\partial f}{\partial \xi^\alpha} \mathrm{d}\xi^\alpha\right) = \frac{\partial^2 f}{\partial \xi^\alpha \partial \xi^\beta} \mathrm{d}\xi^\beta \wedge \mathrm{d}\xi^\alpha \\
&= \frac{1}{2}\left(\frac{\partial^2 f}{\partial \xi^\alpha \partial \xi^\beta} - \frac{\partial^2 f}{\partial \xi^\beta \partial \xi^\alpha}\right) \mathrm{d}\xi^\beta \wedge \mathrm{d}\xi^\alpha = 0.
\end{aligned}$$ (7.22)

In general, $\mathrm{d}\mathrm{d}\theta = 0$ for any k-form (it is actually one of the defining requirements of the exterior derivatives). The forms θ such that $\mathrm{d}\theta = 0$ are called *closed*. Some (but not all) of these forms can be written as $\theta = \mathrm{d}\zeta$, for some other form ζ. Such forms are called *exact*.

Let us illustrate this concept with the example of one-forms. Suppose a one-form is written as $\theta = \theta_\alpha \mathrm{d}\xi^\alpha$. Then, for this form to be exact, $\theta = \mathrm{d}f$, we would need $\theta_\alpha = \frac{\partial f}{\partial \xi^\alpha}$ for some function f. On the other hand, for the form to be closed, we would need

$$\mathrm{d}\theta = \mathrm{d}\theta_\alpha \wedge \mathrm{d}\xi^\alpha = \frac{\partial \theta_\alpha}{\partial \xi^\beta} \mathrm{d}\xi^\beta \wedge \mathrm{d}\xi^\alpha = \frac{1}{2}\left(\frac{\partial \theta_\alpha}{\partial \xi^\beta} - \frac{\partial \theta_\beta}{\partial \xi^\alpha}\right) \mathrm{d}\xi^\beta \wedge \mathrm{d}\xi^\alpha .$$ (7.23)

The requirement $d\theta = 0$ selects a condition on the combination of derivatives of functions $\theta_\alpha(\boldsymbol{\xi})$, which is the n-dimensional generalization of the condition $\mathrm{curl}\mathbf{u} = \mathbf{0}$ for some vector field $\mathbf{u}$. This condition is clearly satisfied if $\theta_\alpha = \frac{\partial f}{\partial \xi^\alpha}$.

Poincaré Lemma If $\boldsymbol{\xi} \in \mathbb{R}^n$, every k-closed form with $k \leq n$, is a derivative of some $k - 1$-form, a result known as *Poincaré lemma*; see [4], p. 133, Proposition 4.2.6 and Exercise 4.2-5. However, if the space is not $\mathbb{R}^n$ and, for example, has "holes," this result no longer holds. For example, we can see that the one-form θ defined on $\mathbb{R}^2 \setminus \mathbf{0}$ is closed:

$$\boldsymbol{\xi} = (x, y), \quad \theta = \frac{x\mathrm{d}y - y\mathrm{d}x}{x^2 + y^2}, \quad d\theta = 0. \tag{7.24}$$

However, this form is not exact. It *looks* like $\theta = \mathrm{d}F$ with $F = \arctan(y/x)$, but one has to remember that $F(x)$ is not defined at $(x, y) = (0, 0)$. To prove that θ is not exact, we compute the integral of the form (Eq. (7.24)) along a circle C of radius r_0 with the origin in the center. Then, in polar coordinates, $x = r_0 \cos\varphi$, $y = r_0 \sin\varphi$, and we compute

$$\theta = \mathrm{d}\varphi, \quad \oint_C \theta = \int_0^{2\pi} \mathrm{d}\varphi = 2\pi \neq 0. \tag{7.25}$$

In contrast, if we had $\theta = \mathrm{d}F$ for some *smooth* function $F(x, y)$ defined everywhere in $\mathbb{R}^2$, we would have $\oint_C \theta = 0$ for any closed contour C. Thus, θ is closed but not exact.

7.2   Pullback of Differential Forms

Suppose a function $\Phi(x)$, $\Phi : X \to Y$ maps a set $X \subset \mathbb{R}^n$ into a set $Y \subset \mathbb{R}^n$. We assume that the derivatives of the function $\Phi(\mathbf{x})$ exist and are sufficiently smooth.

Take any $f(\mathbf{y})$, defined on the set Y. Then, we can define a function with x as the argument through $f(\mathbf{y})$ by simply substituting $\mathbf{y} = \Phi(\mathbf{x})$ into the argument of $f(\mathbf{y})$. We call such function *the pullback* of $f(\mathbf{y})$ by $\Phi(x)$ and denote it $(\Phi^* f)(\mathbf{x})$:

$$(\Phi^* f)(\mathbf{x}) = (f \circ \Phi)(\mathbf{x}) = f(\Phi(\mathbf{x})). \tag{7.26}$$

Note that we don't need to assume anything about $\Phi(x)$ is invertible.

We can also define pullback for differential forms. For example, for one-forms, we have

$$\theta = \sum_\alpha \theta_\alpha(\mathbf{y})\mathrm{d}y^\alpha, \quad \Phi^*\theta = \sum_\alpha \theta_\alpha(\Phi(\mathbf{x}))\mathrm{d}\Phi^\alpha(\mathbf{y}) = \sum_{\alpha,\beta} \theta_\alpha(\Phi(\mathbf{x}))\frac{\partial\Phi^\alpha}{\partial x^\beta}\mathrm{d}x^\beta.$$
$$\tag{7.27}$$

Note that θ is defined in terms of variables $\mathbf{y}$, and $\Phi^*\theta$ is defined in terms of variables $\mathbf{x}$. Similarly, for two-forms, we write

$$\omega = \sum_{\alpha,\beta} \omega_{\alpha\beta}(\mathbf{y})\mathrm{d}y^\alpha \wedge \mathrm{d}y^\beta, \quad \Phi^*\omega = \sum_{\alpha,\beta} \omega_{\alpha\beta}(\Phi(\mathbf{x}))\mathrm{d}\Phi^\alpha(\mathbf{x}) \wedge \mathrm{d}\Phi^\beta(\mathbf{x}), \quad \text{thus}$$

$$\Phi^*\omega = \sum_{\alpha,\beta,p,q} \omega_{\alpha\beta}(\Phi(\mathbf{x}))\frac{\partial \Phi^\alpha}{\partial x^p}(\mathbf{x})\frac{\partial \Phi^\beta}{\partial x^q}(\mathbf{x})\mathrm{d}x^p \wedge \mathrm{d}x^q.$$

$$(7.28)$$

The notion of pullback will be useful in understanding the concept of canonical transformation. Pullbacks of differential forms have the following useful properties:

1. Pullback commutes with the exterior derivative: $\Phi^*\mathrm{d}\alpha = \mathrm{d}\Phi^*\alpha$.
2. Pullback respects the wedge product of forms: $\Phi^*(\alpha \wedge \beta) = (\Phi^*\alpha) \wedge (\Phi^*\beta)$.

We shall leave those two properties without proof.

Pullbacks are useful for dealing with functions and differential forms. We can also define the pushforward, which works naturally for vector fields but is awkward to use for functions and differential forms since they require inverse transformations and their derivatives [4]. For the sake of brevity, we will only consider pullbacks in this book.

7.3 Applications to Hamiltonian Mechanics

Remember the vector field generated by a Hamiltonian X_H defined by Eq. (7.9). We are interested in what happens if this vector field is substituted into the canonical form $\omega = \mathrm{d}q_\alpha \wedge \mathrm{d}p_\alpha = -\mathrm{d}\theta_0$. What we get is a one-form, but which one-form? We need to remember that $\mathrm{d}p_\alpha(X_H)$ selects the $\frac{\partial}{\partial p_\alpha}$ component of X_H in (7.9), so $\mathrm{d}p_\alpha(X_H) = -\frac{\partial H}{\partial q^\alpha}$. Similarly, $\mathrm{d}q_\alpha(X_H) = \frac{\partial H}{\partial p_\alpha}$. Then, for $\omega = \mathrm{d}q^\alpha \wedge \mathrm{d}p_\alpha$,

$$i_{X_H}\omega = i_{X_H}\omega = \mathrm{d}q^\alpha(X_H)\mathrm{d}p_\alpha - \mathrm{d}p_\alpha(X_H)\mathrm{d}q^\alpha = \frac{\partial H}{\partial p_\alpha}\mathrm{d}p_\alpha + \frac{\partial H}{\partial q^\alpha}\mathrm{d}q^\alpha = \mathrm{d}H.$$

$$(7.29)$$

The expression $i_{X_H}\omega = \mathrm{d}H$ is the formulation of Hamiltonian mechanics in the coordinate-free form. It is used extensively in modern research in mechanics since it is true for arbitrary choice of coordinates and is highly useful for writing Hamiltonian mechanics on general manifolds. Similarly, we can take any other function F and state $i_{X_F}\omega = \mathrm{d}F$, when that function is acting as a Hamiltonian.

We can also find an expression for the Poisson bracket by noticing that $X_H(F)$, when X_H is given by Eq. (7.9), is the rate of change $\dot{F}$ of the function $F(\mathbf{q}, \mathbf{p})$ on the solutions of Hamilton's equations (Eq. (6.6)). This rate of change, we remember, is equal to $\{F, H\}$. Thus, we compute

$$\{F, H\} = X_H(F) = \mathrm{d}F(X_H) = i_{X_H}\mathrm{d}F = \omega(X_F, X_H). \qquad (7.30)$$

Thus, the Poisson bracket between any two functions F and G on T^*Q can be defined as the result of the canonical form ω acting on vector fields X_F and X_G. This is also an important result, and it is widely used in modern Hamiltonian mechanics. We will use this result in the next chapter during our discussion of canonical transformations in Chap. 8.

7.4   Integration of Differential Forms: Stokes Formula and Integral Invariants of Poincaré and Poincaré-Cartan

From any one form in the n-dimensional space, we can construct an integral over a one-dimensional curve. We will be interested in the integrals over the closed curves γ, $\oint_\gamma \theta = \oint_\gamma \theta_a(\boldsymbol{\xi})\mathrm{d}\xi^a$. Here, $\gamma(t)$ is a curve in the $2n + 1$-dimensional space $(\mathbf{q}, \mathbf{p}, t)$. We can parameterize the curve by some parameter α and the time evolution of this curve by t.

We have the Stokes or, as Arnol'd puts it [1], the Newton-Leibniz-Gauss-Green-Ostrogradsky-Stokes-Poincaré) formula. For any k-form on a manifold M, and for any surface c (to be more precise, a $k + 1$-chain on M) with the boundary ∂c,

$$\int_c \mathrm{d}\omega = \int_{\partial c} \omega. \qquad (7.31)$$

We shall only use this formula for one-forms integrated over one-dimensional contours and two-forms integrated over surfaces in the $2n$ or $2n + 1$-dimensional manifolds, depending on whether we take into account the time or not.

Consider the $2n + 1$-dimensional space, also called the *extended phase space*, consisting of coordinates $(\mathbf{q}, \mathbf{p}, t)$. Suppose we take a contour γ_0 in that space defined by the curve $\gamma_0(\alpha)$ in the space $(\mathbf{q}_0(\alpha), \mathbf{p}_0(\alpha), t_0(\alpha))$. Let this contour be advected by the Hamiltonian flow; in other words, at every time, every coordinate $(\mathbf{q}(\alpha, t), \mathbf{p}(\alpha, t), t)$ is the solution of Hamilton's equation with the Hamiltonian H and initial conditions $(\mathbf{q}_0(\alpha), \mathbf{p}_0(\alpha), t_0(\alpha))$. The initial curve moves in space and creates a two-dimensional tube, parameterized by α and t. Such contours are called *vortex tubes*. The concept of vortex tubes is illustrated in Fig. 7.1.

A theorem [1], Sect. 44, shows that the integral of a particular one-form along any two curves Γ_1 and Γ_2 encircling the vortex tube is the same:

$$\oint_{\Gamma_1} \theta = \oint_{\Gamma_2} \theta, \quad \text{where} \quad \theta := \mathbf{p} \cdot \mathrm{d}\mathbf{q} - H\mathrm{d}t. \qquad (7.32)$$

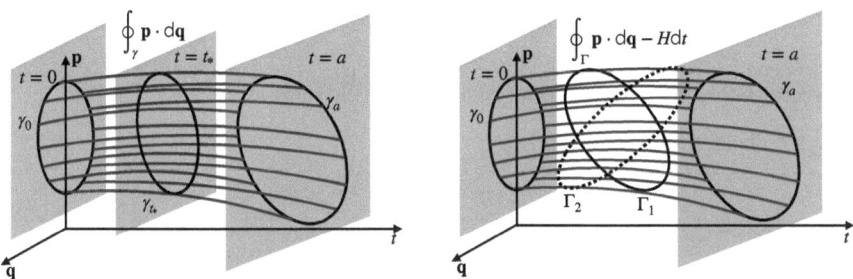

Fig. 7.1 Illustration of vortex tubes and integral invariants. We start with a contour γ_0 in $(\mathbf{q}, \mathbf{p})$ plane for a fixed t and advect that contour by solutions of Hamilton's equations, creating a vortex tube. On the left panel, we show Poincare's invariant, which computes the integral $\oint_{\gamma(t)} \mathbf{p} \cdot d\mathbf{q}$ for a fixed t. On the right panel, we show Poincar'e-Cartan's integral invariant, which is the integral of the one-form $\theta = \mathbf{p} \cdot d\mathbf{q} - H dt$ over any closed contour Γ lying the vortex tube according to (7.32)

The form θ defined by (7.32) is called *the integral invariant of Poincaré-Cartan*. As a corollary, if we restrict the contours to be only at the lines $t =$const, then we get the conservation of *relative integral invariant of Poincaré*:

$$\oint_{\gamma_0} \mathbf{p} \cdot d\mathbf{q} = \oint_{\gamma(t_*)} \mathbf{p} \cdot d\mathbf{q}, \tag{7.33}$$

where $\gamma(t_*)$ is the curve that is obtained from the initial curve γ_0 with $t_* =$const. Integral invariants of Poincaré and Poincaré-Cartan are illustrated on the left and the right panels of Fig. 7.1, respectively. The integral invariants described here will be important for analyzing canonical transformations in the next section.

7.5 Looking Forward

Differential forms, vector fields, and, in general, the concepts of differential geometry and tensor analysis underpin all modern mechanics. A reader is encouraged to reach out to classical texts such as Marsden [51], José and Saletan [5], Abraham and Marsden [50], Holm [2, 3], Marsden and Ratiu [4], Arnol'd [1], and many other excellent books with detailed exposition of mathematical concepts forming the foundation of all modern mechanics. Understanding these concepts is essential for anyone interested in continuing their studies in modern theory and applications of mechanics.

7.6   Practice Problems

1. Vector fields, differential forms, and external calculus:

   (a) Consider the vector fields

   $$X = \xi^1 \frac{\partial}{\partial \xi^1} + \xi^2 \frac{\partial}{\partial \xi^2} + \xi^3 \frac{\partial}{\partial \xi^3}, \quad Y = -\xi^2 \frac{\partial}{\partial \xi^1} + \xi^1 \frac{\partial}{\partial \xi^2} + (\xi^1 + \xi^2) \frac{\partial}{\partial \xi^3}.$$

   Find $[X, Y]$.

   (b) Suppose $\boldsymbol{\xi} \in \mathbb{R}^3$ and $f(\boldsymbol{\xi}) = (\xi^1 + \xi^2)^2 + (\xi^3)^2$. Find df, $i_X df$, and $i_Y df$.

   (c) Suppose $\boldsymbol{\xi} \in \mathbb{R}^3$ and $\theta = \xi^1 d\xi^1 + (\xi^2 - \xi^3)d\xi^2 + (\xi^3 - \xi^2)d\xi^3$. Find $d\theta$. Is θ exact? Is it closed?

   (d) $\boldsymbol{\xi} \in \mathbb{R}^3$ and $\theta = \xi^3 d\xi^1 + (\xi^2 - \xi^1)d\xi^2 + (\xi^3 - \xi^2)d\xi^3$. Given (X, Y) from a), find $i_X \theta$, $i_Y i_X d\theta$, and $di_X \theta$.

2. Vector calculus and differential forms:

   (a) (Curl) Consider $\boldsymbol{\xi} \in \mathbb{R}^3$, and one-form $\theta = v_\alpha(\boldsymbol{\xi})d\xi^\alpha$. Compute $d\theta$, and compare the components of $d\xi^2 \wedge d\xi^3$, $d\xi^3 \wedge d\xi^1$, $d\xi^1 \wedge d\xi^2$, with the components of $\text{curl}(v_1, v_2, v_3)$.

   (b) If θ above is obtained as $\theta = dF$ for some scalar function F, what vector identity corresponds to the exterior calculus identity $ddF = 0$?

   (c) (div) Consider $\boldsymbol{\xi} \in \mathbb{R}^3$, and two-form

   $$\theta = v_1(\boldsymbol{\xi})d\xi^2 \wedge d\xi^3 + v_2(\boldsymbol{\xi})d\xi^3 \wedge d\xi^1 + v_3(\boldsymbol{\xi})d\xi^1 \wedge d\xi^2.$$

   Compute $d\theta$.

3. Suppose θ is one-form, and X and Y are arbitrary vector fields. Show that the coordinate-free definition of the exterior derivative

   $$i_Y i_X d\theta = d\theta(X, Y) = i_X d(i_Y \theta) - i_Y d(i_X \theta) - i_{[X,Y]}\theta \tag{7.34}$$

   coincides with coordinate-wise definition (Eq. (7.19)) in local coordinates. In other words, compute both sides of Eq. (7.34) in local coordinates and show that they are equal to each other.

Chapter 8
Canonical Transformations

8.1 General Properties

Our goal here is to make Hamilton's equations simpler by transforming the coordinates $(\mathbf{q}, \mathbf{p})$ to some coordinates $(\mathbf{P}, \mathbf{Q})$, when the equations will, hopefully, become simpler and yield at least a partial solution to the equations, for example, conservation laws. We want to keep the structure of canonical equations (Eq. (6.6)) in the new coordinates, *i.e.*, there should be some new Hamiltonian $K(\mathbf{Q}, \mathbf{P}, t)$ such that the system in the new variables is

$$\frac{d\mathbf{Q}}{dt} = \frac{\partial K}{\partial \mathbf{P}}, \quad \frac{d\mathbf{P}}{dt} = -\frac{\partial K}{\partial \mathbf{Q}}. \tag{8.1}$$

Suppose $(\mathbf{Q}(\mathbf{q}, \mathbf{p}), \mathbf{P}(\mathbf{q}, \mathbf{p}))$ are transformations of coordinates and momenta. We call that transformation for shortness $\Phi(\mathbf{q}, \mathbf{p})$. Take any two functions $F(\mathbf{Q}, \mathbf{P})$, $G(\mathbf{Q}, \mathbf{P})$, and define the new functions:

$$f(\mathbf{q}, \mathbf{p}) = \Phi^* F = F\big(\mathbf{Q}(\mathbf{q}, \mathbf{p}), \mathbf{P}(\mathbf{q}, \mathbf{p})\big) = F \circ \Phi,$$
$$g(\mathbf{q}, \mathbf{p}) = \Phi^* G = G\big(\mathbf{Q}(\mathbf{q}, \mathbf{p}), \mathbf{P}(\mathbf{q}, \mathbf{p})\big) = G \circ \Phi. \tag{8.2}$$

We remind the reader that the composition $\circ \, \Phi$ in Eq. (8.2) simply means that you substitute the components of $\Phi = (\mathbf{Q}(\mathbf{q}, \mathbf{p}), \mathbf{P}(\mathbf{q}, \mathbf{p}))$ into the arguments of $F(\mathbf{Q}, \mathbf{P})$ and $G(\mathbf{Q}, \mathbf{P})$.

Define $\{f, g\}_{(\mathbf{q},\mathbf{p})}$ to be the canonical Poisson bracket in the $(\mathbf{q}, \mathbf{p})$ coordinates, and $\{F, G\}_{(\mathbf{Q},\mathbf{P})}$ the corresponding bracket in the $(\mathbf{Q}, \mathbf{P})$ coordinates:

$$\{f, g\}_{(\mathbf{q},\mathbf{p})} = \frac{\partial f}{\partial \mathbf{q}} \cdot \frac{\partial g}{\partial \mathbf{p}} - \frac{\partial f}{\partial \mathbf{p}} \cdot \frac{\partial g}{\partial \mathbf{q}}, \quad \{F, G\}_{(\mathbf{Q},\mathbf{P})} = \frac{\partial F}{\partial \mathbf{Q}} \cdot \frac{\partial G}{\partial \mathbf{P}} - \frac{\partial F}{\partial \mathbf{P}} \cdot \frac{\partial G}{\partial \mathbf{Q}}. \tag{8.3}$$

© The Author(s), under exclusive license to Springer Nature Switzerland AG 2025
V. Putkaradze, *A Concise Introduction to Classical Mechanics*,
Surveys and Tutorials in the Applied Mathematical Sciences 16,
https://doi.org/10.1007/978-3-031-84977-0_8

There are several definitions of canonical transformations you may encounter in the literature.

1. **Definition 1** A canonical transformation preserves the canonical structure of equations (Eq. (8.1)), so equations in the new variables $(\mathbf{Q}, \mathbf{P})$ have the form (Eq. (8.1)); see [6].
2. **Definition 2** A canonical transformation preserves the canonical Poisson bracket $\{\cdot, \cdot\}$. More precisely, if $F(\mathbf{Q}, \mathbf{P})$ and $G(\mathbf{Q}, \mathbf{P})$ are two functions, and $f(\mathbf{q}, \mathbf{p}) = F \circ \Phi$, and $g(\mathbf{q}, \mathbf{p}) = G \circ \Phi$, then

$$\{f, g\}_{(\mathbf{q}, \mathbf{p})} = \{F \circ \Phi, G \circ \Phi\}_{(\mathbf{q}, \mathbf{p})} = \{F, G\}_{(\mathbf{Q}, \mathbf{P})} \circ \Phi. \tag{8.4}$$

3. **Definition 3** A canonical two-form $\omega = \sum_{\alpha} dp^{\alpha} \wedge dq^{\alpha}$ is preserved under the transformation [1]:

$$\sum_{\alpha} dQ^{\alpha} \wedge dP_{\alpha} = \sum_{\alpha} dq^{\alpha} \wedge dp_{\alpha}. \tag{8.5}$$

These three definitions are actually *not* equivalent, as pointed out in [1], note on p. 241. Namely, for the first definition to be true, one can prove that the conservation of the Poisson bracket could be written in a more general form; see [52], Chap. 11.4, p. 341:

$$\{f, g\}_{(\mathbf{q}, \mathbf{p})} = \{F \circ \Phi, G \circ \Phi\}_{(\mathbf{q}, \mathbf{p})} = a\{F, G\}_{(\mathbf{Q}, \mathbf{P})} \circ \Phi. \tag{8.6}$$

Thus, if you use Definition 1, you need to be cognizant of the fact that it is not equivalent to the more standard Definitions 2 and 3. The constant a in (8.6) is called the *valence* of the canonical transformation. Transformations with $a = 1$ are called *univalent*; they play a special role in the theory of canonical transformation. Definitions 2 and 3 refer to univalent transformations. In this book, we only consider univalent transformations as they are particularly important for mechanics. You will see non-univalent transformations in many classical books on analytical mechanics.

We start by assuming that Definition 2 is true, so the Poisson bracket is conserved according to Eq. (8.4). In that case, the equation of motion in the new coordinates is going to be canonical, so Eq. (8.1) is also true. We want to eventually prove that Definition 3 is true as well, and it is equivalent to Definition 2, but we will need some preparatory work for that.

First, using Eq. (8.4), we will prove an important result showing the relationship between gradients of transformations. This result will be very useful for further consideration of Hamiltonian mechanics.

Lemma 8.1 (On Symplecticity of Canonical Transformations) *Suppose the transformation* $(\mathbf{Q}, \mathbf{P}) = \Phi(\mathbf{q}, \mathbf{p}, t)$ *preserves the Poisson bracket, so Eq. (8.4) is* *satisfied. Denote* $\mathbf{u} = (\mathbf{q}, \mathbf{p})$. *Then, the $2n \times 2n$ matrix of gradients $\nabla_{\mathbf{u}}\Phi$ satisfies*

$$\left(\frac{\partial \Phi}{\partial \mathbf{u}}\right)^T \mathbb{J} \frac{\partial \Phi}{\partial \mathbf{u}} = \mathbb{J}, \quad \mathbb{J} := \begin{pmatrix} \mathbf{0}_{n\times n} & \mathbb{I}_{n\times n} \\ -\mathbb{I}_{n\times n} & \mathbf{0}_{n\times n} \end{pmatrix}. \tag{8.7}$$

Transformations $\Phi(\mathbf{u})$ satisfying the derivative condition (Eq. (8.7)) are called symplectic.

Proof The Poisson bracket in the canonical variables $(\mathbf{Q}, \mathbf{P})$ for two functions $F(\mathbf{Q}, \mathbf{P})$ and $G(\mathbf{Q}, \mathbf{P})$ can be alternatively written in the form

$$\{F, G\}_{(\mathbf{Q}, \mathbf{P})} = (\nabla_{\mathbf{U}} F(\mathbf{Q}, \mathbf{P})) \cdot \mathbb{J} (\nabla_{\mathbf{U}} F(\mathbf{Q}, \mathbf{P})), \quad \text{where} \quad \mathbf{U} = (\mathbf{Q}, \mathbf{P}). \tag{8.8}$$

Take $f(\mathbf{q}, \mathbf{p}) = F \circ \Phi$, $g(\mathbf{q}, \mathbf{p}) = G \circ \Phi$. Then, from Eq. (8.8) and from the fact that the motion in the coordinates $(\mathbf{q}, \mathbf{p})$ is canonical, we obtain

$$\begin{aligned}
\{f, g\}_{(\mathbf{q}, \mathbf{p})} &= (\nabla_{\mathbf{u}} (F \circ \Phi)) \cdot \mathbb{J} (\nabla_{\mathbf{u}} (G \circ \Phi)) \\
&= \left(\frac{\partial \Phi}{\partial \mathbf{u}} \cdot (\nabla_{\mathbf{U}} F) \circ \Phi)\right) \cdot \mathbb{J} \left(\frac{\partial \Phi}{\partial \mathbf{u}} \cdot (\nabla_{\mathbf{U}} G) \circ \Phi\right) \\
&= ((\nabla_{\mathbf{U}} F) \circ \Phi) \left(\left(\frac{\partial \Phi}{\partial \mathbf{u}}\right)^T \mathbb{J} \frac{\partial \Phi}{\partial \mathbf{u}}\right) ((\nabla_{\mathbf{U}} G) \circ \Phi).
\end{aligned} \tag{8.9}$$

Because the transformation Φ preserves the Poisson bracket according to Eq. (8.4), we also have

$$\{f, g\}_{(\mathbf{q}, \mathbf{p})} = \{F, G\}_{(\mathbf{Q}, \mathbf{P})} \circ \Phi = ((\nabla_{\mathbf{U}} F) \circ \Phi) \mathbb{J} ((\nabla_{\mathbf{U}} G) \circ \Phi), \tag{8.10}$$

which is only possible if Eq. (8.7) is satisfied. ∎

Corollary 8.2 (Non-degeneracy of Canonical Transformations) *Transformations $\Phi(\mathbf{q}, \mathbf{p}, t)$ preserving the Poisson bracket as given by (8.4) are invertible; the inverse transformation $\Psi(\mathbf{Q}, \mathbf{P}, t) = \Phi^{-1}(\mathbf{Q}, \mathbf{P}, t)$ exists and also preserves the Poisson bracket.*

Proof Taking the determinant of both sides of Eq. (8.7), we see that $(\det \nabla_{\mathbf{u}} \Phi)^2 = 1 \neq 0$, so the matrix $\nabla_{\mathbf{u}} \Phi$ is non-singular. Thus, by the implicit function theorem, the inverse transformation $\Psi = \Phi^{-1}$ exists and is also differentiable. Composing both sides of Eq. (8.4) with Ψ, we get

$$\{f, g\}_{(\mathbf{q}, \mathbf{p})} \circ \Psi = \{f \circ \Psi, g \circ \Psi\}_{(\mathbf{Q}, \mathbf{P})},$$

so $\Psi(\mathbf{Q}, \mathbf{P})$ also preserves the Poisson bracket. ∎

We can now prove that Definitions 2 and 3 are equivalent.

Theorem 8.3 (On the Conservation of Canonical Two-Form) *The transformation* $\Phi(\mathbf{q}, \mathbf{p}, t)$ *mapping* $(\mathbf{q}, \mathbf{p})$ *to* $(\mathbf{P}, \mathbf{Q})$ *is canonical as defined by Eq. (8.4) if and only if the canonical two-form is preserved under the transformation according to Eq. (8.5).*

Note The condition of the preservation of canonical two-form is often written as Eq. (8.5). Of course, this condition needs to be formulated in terms of the same variables. Hence, a mathematically rigorous way of formulating Eq. (8.5) is through the concept of pullback of the two-form as defined in Eq. (7.28):

$$\Phi^* \sum_\alpha dQ^\alpha \wedge dP_\alpha = \sum_\alpha dq^\alpha \wedge dp_\alpha . \tag{8.11}$$

Note also that the differential dQ^α and dP_α in Eq. (8.5) only refers to the phase space variables $(\mathbf{q}, \mathbf{p})$ and does not contain derivatives with respect to t, which is made natural through the definition of a pullback in Eq. (8.11), treating time t as a parameter.

Proof of Theorem (8.3)
Part 1: Remember that the Poisson bracket can be defined through the canonical two-form Eq. (7.30), which we remind here: $\{F, H\} = \omega(X_F, X_H)$. Thus, the conservation of the canonical two-form (Eq. (8.5)) gives the preservation of the Poisson bracket, and hence the transformation is canonical according to Eq. (8.4).
Part 2: Suppose the transformation is canonical according to Eq. (8.4). For the canonical two-form in the new coordinates, we write

$$\Phi^* \sum_\alpha dQ^\alpha \wedge dP_\alpha = \sum_{\alpha,\beta,\gamma} \frac{\partial Q^\alpha}{\partial q^\beta} \frac{\partial P_\alpha}{\partial q^\gamma} dq^\beta \wedge dq^\gamma + \sum_{\alpha,\beta,\gamma} \frac{\partial Q^\alpha}{\partial p^\beta} \frac{\partial P_\alpha}{\partial p^\gamma} dp^\beta \wedge dp^\gamma$$
$$+ \sum_{\alpha,\beta,\gamma} \left(\frac{\partial Q^\alpha}{\partial q^\beta} \frac{\partial P_\alpha}{\partial p^\gamma} + \frac{\partial Q^\alpha}{\partial p^\beta} \frac{\partial P_\alpha}{\partial p_\gamma} \right) dq^\beta \wedge dp_\gamma . \tag{8.12}$$

We want to prove that the complicated-looking expression in Eq. (8.12) is equal to the canonical two-form in the original coordinates $\sum_\alpha dq^\alpha \wedge dp_\alpha$.

We know that according to Corollary 8.2, the inverse transformation exists and is also canonical. Let us compute the following derivative using Poisson brackets in the new and old variables:

$$\frac{\partial P_\alpha}{\partial q^\beta} = \left\{ P_\alpha(\mathbf{q}, \mathbf{p}, t), p_\beta \right\}_{(\mathbf{q}, \mathbf{p})} = \left\{ P_\alpha, p_\beta(\mathbf{Q}, \mathbf{P}, t) \right\}_{(\mathbf{Q}, \mathbf{P})} \circ \Phi = -\frac{\partial p_\beta}{\partial Q^\alpha} \circ \Phi . \tag{8.13}$$

Proceeding similarly, we obtain for all four cases of derivatives:

$$\frac{\partial P_\alpha}{\partial q^\beta} = -\frac{\partial p_\beta}{\partial Q^\alpha} \circ \Phi, \quad \frac{\partial P_\alpha}{\partial p_\beta} = \frac{\partial q^\beta}{\partial Q^\alpha} \circ \Phi,$$

$$\frac{\partial Q^\alpha}{\partial p_\beta} = -\frac{\partial q^\beta}{\partial P_\alpha} \circ \Phi, \quad \frac{\partial Q^\alpha}{\partial q^\beta} = \frac{\partial p_\beta}{\partial P_\alpha} \circ \Phi.$$

(8.14)

Then, using Eq. (8.14) and the fact that the Poisson bracket is conserved according to Eq. (8.4), the new canonical two-form (Eq. (8.12)) is written as

$$\Phi^* \sum_\alpha dQ^\alpha \wedge dP_\alpha = \sum_{\beta,\gamma} \{p_\beta, p_\gamma\} dq^\beta \wedge dq^\gamma$$

(8.15)

$$+ \{q^\gamma, p_\beta\} dq^\beta \wedge dp_\gamma + \{q^\beta, q^\gamma\} dp_\beta \wedge dp_\gamma.$$

Using the standard relationship for the Poisson bracket $\{p_\beta, p_\gamma\} = 0$, $\{q^\beta, q^\gamma\} = 0$, and $\{q^\gamma, p_\beta\} = \delta^\gamma_\beta$, we obtain that $\sum_\alpha dQ^\alpha \wedge dP_\alpha = \sum_\alpha dq^\alpha \wedge dp_\alpha$, so the canonical two-form is preserved under canonical transformations. ∎

The Necessary and Sufficient Condition for Canonical Transformation The condition (Eq. (8.4)) can be written in terms of coordinates as

$$\{Q^\alpha, Q^\beta\}_{(\mathbf{q},\mathbf{p})} = 0, \quad \{P_\alpha, P_\beta\}_{(\mathbf{q},\mathbf{p})} = 0, \quad \{Q^\alpha, P_\beta\}_{(\mathbf{q},\mathbf{p})} = \delta^\alpha_\beta.$$

(8.16)

The Poisson bracket in Eq. (8.16) is taken with respect to coordinates $(\mathbf{q}, \mathbf{p})$. An equivalent condition can be derived to require that the inverse transformations $\mathbf{q} = \mathbf{q}(\mathbf{Q}, \mathbf{P})$ and $\mathbf{p} = \mathbf{p}(\mathbf{Q}, \mathbf{P})$ are canonical. The condition (Eq. (8.16)) gives a straightforward way of verifying whether a given transformation is canonical.

Let us now turn our attention to the time-independent canonical transformations. As it turns out, these transformations preserve not just the Poisson bracket but also the Hamiltonian.

Theorem 8.4 (Time-Independent Canonical Transformations) *Suppose* $(\mathbf{q}, \mathbf{p})$ *satisfy the canonical equations:*

$$\frac{d\mathbf{q}}{dt} = \frac{\partial h}{\partial \mathbf{p}}, \quad \frac{d\mathbf{q}}{dt} = -\frac{\partial h}{\partial \mathbf{q}}, \quad h = h(\mathbf{q}, \mathbf{p}, t).$$

(8.17)

Time-independent canonical transformation $\Phi : (\mathbf{q}, \mathbf{p}) \rightarrow (\mathbf{P}, \mathbf{Q})$ *preserves the structure of canonical equations in the new variables (Eq. (8.1)) with* $K = H = h \circ \Phi^{-1}(\mathbf{Q}, \mathbf{P})$.

Proof Suppose Eq. (8.4) is true and the evolution for $(\mathbf{q}, \mathbf{p})$ is given by canonical equations (Eq. (8.17)). According to Corollary 8.2, the inverse transformation $\Psi(\mathbf{Q}, \mathbf{P}) = \Phi^{-1}(\mathbf{Q}, \mathbf{P})$ exists and also preserves the Poisson bracket. We need to prove that the evolution equations for $(\mathbf{Q}, \mathbf{P})$ are also given by the canonical equations (Eq. (8.1)) for $K(\mathbf{Q}, \mathbf{P}) = h \circ \Psi(\mathbf{Q}, \mathbf{P})$. Then, for each $\alpha = 1, \ldots, n$, and for any functions $Q^\alpha(\mathbf{q}, \mathbf{p})$ and $P^\alpha(\mathbf{q}, \mathbf{p})$, and for any Hamiltonian $h(\mathbf{q}, \mathbf{p})$, we have

$$\frac{d}{dt}Q^\alpha(\mathbf{q}, \mathbf{p}) = \{Q^\alpha, h\}_{(\mathbf{q}, \mathbf{p})}, \quad \frac{d}{dt}P_\alpha(\mathbf{q}, \mathbf{p}) = \{P_\alpha, h\}_{(\mathbf{q}, \mathbf{p})}. \tag{8.18}$$

Compose these evolution equations (Eq. (8.18)) with $\Phi(\mathbf{Q}, \mathbf{P})$, and remember that $\mathbf{Q}(\mathbf{q}, \mathbf{p}) \circ \Psi = \mathbf{Q}$, $\mathbf{P}(\mathbf{q}, \mathbf{p}) \circ \Psi = \mathbf{P}$. Using Eq. (8.4), we obtain

$$\frac{dQ^\alpha}{dt} = \{Q^\alpha, H\}_{(\mathbf{Q}, \mathbf{P})} = \frac{\partial H}{\partial P_\alpha}, \quad \frac{dP_\alpha}{dt} = \{P_\alpha, H\}_{(\mathbf{Q}, \mathbf{P})} = -\frac{\partial H}{\partial Q_\alpha}, \tag{8.19}$$

so the new Hamiltonian is $K = H(\mathbf{Q}, \mathbf{P}, t) = h(\mathbf{q}, \mathbf{p}, t) \circ \Psi(\mathbf{Q}, \mathbf{P})$. ∎

Note One can take $K = H + k(t)$, where $k(t)$ is an arbitrary function of time; in fact, some references do that. We will ignore that extra time dependence as it plays no role in the dynamics.

Examples Let us consider some examples of $Q = Q(q, p)$ and $P = P(q, p)$ for a one-dimensional system. For these cases, in condition (Eq. (8.16)) $\{Q, Q\} = 0$ and $\{P, P\} = 0$, so we only need to check $\{Q, P\} = a$.

1. $Q = pe^q$, $P = p + q$. Then,

$$\{Q, P\} = \frac{\partial Q}{\partial q}\frac{\partial P}{\partial p} - \frac{\partial Q}{\partial p}\frac{\partial P}{\partial q} = pe^q - e^q = (p - 1)e^q \neq 1.$$

The transformation is not canonical.
2. $Q = p + q$, $P = \frac{1}{2}(p - q)$. Similar to the previous case, $\{Q, Q\} = 0$ and $\{P, P\} = 0$. We check

$$\{Q, P\} = \frac{\partial Q}{\partial q}\frac{\partial P}{\partial p} - \frac{\partial Q}{\partial p}\frac{\partial P}{\partial q} = \frac{1}{2} + \frac{1}{2} = 1.$$

The transformation is canonical.

8.2   Generating Functions of Canonical Transformations

Let us now see how to derive expressions for canonical transformations using the concept of generating function. The idea is that we specify a function, called *generating function*, and compute the coordinate transformations for some coordinates using the derivatives of this function. The transformations themselves are then obtained by inverting some of the functional relationships. This is the topic where our knowledge of differential forms we learned in Chap. 7 will be useful.

Suppose such a transformation from $(\mathbf{q}, \mathbf{p})$ to $(\mathbf{Q}, \mathbf{P})$ is canonical. To derive the expression for the new Hamiltonian K, remember the conservation of the integral invariant of Poincaré-Cartan (Eq. (7.32)). If that integral has to be preserved in the new coordinates $(\mathbf{Q}, \mathbf{P}, t)$, the forms $\mathbf{p} \cdot d\mathbf{q} - H dt$ and $\mathbf{P} \cdot d\mathbf{Q} - K dt$ should differ by an exact form in $(\mathbf{q}, \mathbf{p}, t)$ space, which we write as dF, so their integral over any closed contour tracing the flow tube should be exactly the same. We thus obtain

$$\mathbf{p} \cdot d\mathbf{q} - H dt = \mathbf{P} \cdot d\mathbf{Q} - K dt + dF . \tag{8.20}$$

In our case, one can prove the following:

Theorem 8.5 *Transformations satisfying Eq. (8.20) leads to canonical equations Suppose $(\mathbf{q}, \mathbf{p})$ are satisfying canonical equations with the Hamiltonian $H(\mathbf{q}, \mathbf{p}, t)$; the functions $\mathbf{P}(\mathbf{q}, \mathbf{p}, t)$, $\mathbf{Q}(\mathbf{q}, \mathbf{p}, t)$ are new coordinates in the extended phase space, and $K(\mathbf{P}, \mathbf{Q}, t)$ and $F(\mathbf{Q}, \mathbf{P}, t)$ is such that Eq. (8.20) holds. Then, the trajectories in coordinates $(\mathbf{Q}(t), \mathbf{P}(t))$ in the phase space satisfy canonical equations (Eq. (8.1)) with K being the new Hamiltonian.*

This result can be proven even in a more general case of when the time is also transformed; see the proof of this result in [1], Chap. 45.

The function $F(\mathbf{Q}, \mathbf{P}, t)$ is called the generating function of the canonical transformation. It is hard to find canonical transformations $\mathbf{P}(\mathbf{p}, \mathbf{q}, t)$ and $\mathbf{Q}(\mathbf{p}, \mathbf{q}, t)$ for a given F from Eq. (8.20) as they, in general, involve solving some nonlinear partial differential equations. There are other ways of writing the analogs of Eq. (8.20), where only some parts of the independent variables are lower case ("old") and other parts are upper case ("new"). Four typical cases are considered below. In these four cases, the function F can be taken as an arbitrary function of the arguments, and the expressions for the canonical transformation follow. The payback for this apparent simplicity of using the generating function will be the necessity to find the inverse of the transformations to bring them to the form $(\mathbf{Q}, \mathbf{P})$ as functions of $(\mathbf{q}, \mathbf{p}, t)$. The new Hamiltonian K will be expressed from the analogs of Eq. (8.20) by collecting the terms proportional to dt.

There are four standard types of generating functions, although one could also design other types by mixing up different versions of old and new variables. The four standard generating functions for canonical transformations are given below. For us to be able to invert the equations and compute $(\mathbf{Q}, \mathbf{P})$ as a function of $(\mathbf{q}, \mathbf{p})$, the generating functions $F_i, i = 1, \ldots 4$ defined below have to satisfy a non-degeneracy condition on the matrix of second derivatives of F. We assume that these conditions are satisfied and all inverse transformations can be found.

Type 1: $(\mathbf{q}, \mathbf{Q})$ as Independent Variables In Eq. (8.20), the differentials are already written in terms of dq^α and dQ^α. This is perhaps the most popular type of canonical transformation. The generating function is $F = F_1(\mathbf{q}, \mathbf{Q}, t)$. The equations for $(\mathbf{p}, \mathbf{P})$ expressed in terms of the generating function F_1 are

$$
\begin{cases}
p_\alpha = \dfrac{\partial F_1}{\partial q^\alpha} \\[2ex]
P_\alpha = -\dfrac{\partial F_1}{\partial Q^\alpha}
\end{cases}
\tag{8.21}
$$

Since $\mathbf{q}$ and $\mathbf{Q}$ are independent variables, the new Hamiltonian K obtained by considering the term proportional to dt in Eq. (8.20) takes the following form:

$$
K = H + \frac{\partial F_1}{\partial t} .
\tag{8.22}
$$

Non-degeneracy Conditions To obtain the actual transformation, we need to invert either the equation for $\mathbf{p}$ to find $\mathbf{Q} = \mathbf{Q}(\mathbf{q}, \mathbf{p}, t)$ or the equation for $\mathbf{P}$ to find the equation for $\mathbf{p} = \mathbf{p}(\mathbf{Q}, \mathbf{P}, t)$. In both cases, the equations may be inverted if

$$
\det\left(\frac{\partial^2 F_1}{\partial \mathbf{Q} \partial \mathbf{q}}\right) =
\begin{vmatrix}
\dfrac{\partial^2 F_1}{\partial Q_1 \partial q_1} & \cdots & \dfrac{\partial^2 F_1}{\partial Q_1 \partial q_n} \\[2ex]
\cdots & \cdots & \cdots \\[2ex]
\dfrac{\partial^2 F_1}{\partial Q_n \partial q_1} & \cdots & \dfrac{\partial^2 F_1}{\partial Q_n \partial q_n}
\end{vmatrix}
\neq 0 .
\tag{8.23}
$$

Thus, for example, we cannot take the generating function to be $F_1 = A(\mathbf{q}) + B(\mathbf{Q})$, as such generating function will not lead to a meaningful canonical transformation. Similar non-degeneracy conditions can be readily derived for the generating functions of types 2, 3, and 4 below.

Type 2: $(\mathbf{q}, \mathbf{P})$ as Independent Variables First, let us change the definition of F in Eq. (8.20) as

$$
F = F_2 - \mathbf{P} \cdot \mathbf{Q}(\mathbf{q}, \mathbf{P}, t)
\tag{8.24}
$$

Note that the change of generating functions from F to F_2 involves the unknown function $\mathbf{Q}(\mathbf{q}, \mathbf{P}, t)$. The differentials are taken only with respect to the variables and not time, except for the function F. This change of variables rewrites Eq. (8.20) only with differentials of the independent variables:

$$
\mathbf{p} \cdot d\mathbf{q} - H dt = -\mathbf{Q} \cdot d\mathbf{P} - K dt + dF_2.
\tag{8.25}
$$

We obtain the following equations for the unknown variables $(\mathbf{Q}, \mathbf{p})$:

$$
\begin{cases}
p_\alpha = \dfrac{\partial F_2}{\partial q^\alpha} \\[2ex]
Q_\alpha = \dfrac{\partial F_2}{\partial P_\alpha}
\end{cases}
\tag{8.26}
$$

The new Hamiltonian is then obtained by collecting the terms proportional to dt in Eq. (8.25):

$$K = H + \frac{\partial F_2}{\partial t}. \tag{8.27}$$

Type 3: $(\mathbf{p}, \mathbf{Q})$ as Independent Variables We need to express $\mathbf{q} = \mathbf{q}(\mathbf{p}, \mathbf{Q}, t)$, $F = F(\mathbf{p}, \mathbf{Q}, t)$ and take derivatives in Eq. (8.20). We change the definition of the generating function F as We take the generating function $F = F_3(\mathbf{p}, \mathbf{Q}, t) + \mathbf{q}(\mathbf{p}, \mathbf{Q}, t) \cdot \mathbf{p}$. Equation (8.20) then becomes

$$- \mathbf{q} \cdot d\mathbf{p} - H dt = \mathbf{P} \cdot d\mathbf{Q} - K dt + dF_3. \tag{8.28}$$

Collecting the terms proportional to $d\mathbf{p}$ and $d\mathbf{Q}$ gives equations for the unknown functions $(\mathbf{P}, \mathbf{q})$:

$$\begin{cases} q^\alpha = -\dfrac{\partial F_3}{\partial p_\alpha} \\[3mm] P_\alpha = -\dfrac{\partial F_3}{\partial Q^\alpha} \end{cases} \tag{8.29}$$

The new Hamiltonian is readily obtained from Eq. (8.28) by collecting terms proportional to dt as

$$K = H + \frac{\partial F_3}{\partial t}. \tag{8.30}$$

Type 4: $(\mathbf{p}, \mathbf{P})$ as Independent Variables In that case, we need to express $\mathbf{q} = \mathbf{q}(\mathbf{p}, \mathbf{P}, t)$, $\mathbf{Q} = \mathbf{Q}(\mathbf{p}, \mathbf{P}, t)$, $F = F(\mathbf{p}, \mathbf{P}, t)$. We take $F = F_4(\mathbf{p}, \mathbf{P}, t) + \mathbf{q}(\mathbf{p}, \mathbf{P}, t) \cdot \mathbf{p} - \mathbf{Q}(\mathbf{p}, \mathbf{P}, t) \cdot \mathbf{P}$. and substitute in Eq. (8.20). We obtain

$$- \mathbf{q} \cdot d\mathbf{p} - H dt = -\mathbf{Q} \cdot d\mathbf{P} - K dt + dF_4. \tag{8.31}$$

Collecting the terms proportional to $(\mathbf{p}, \mathbf{P})$, we obtain equations for coordinate transformations:

$$\begin{cases} q^\alpha = -\dfrac{\partial F_4}{\partial p_\alpha} \\[3mm] Q^\alpha = \dfrac{\partial F_4}{\partial P_\alpha} \end{cases} \tag{8.32}$$

The new Hamiltonian is computed from Eq. (8.31) in the familiar form:

$$K = H + \frac{\partial F_4}{\partial t}. \tag{8.33}$$

Example: Type 1 Transformation Consider an n-dimensional system, and $\mathbf{p} = (p_1, \ldots, p_n)$, $\mathbf{q} = (q^1, \ldots, q^n)$, *etc.* Consider the generating function F_1 and the Hamiltonian $H(\mathbf{q}, \mathbf{p}, t)$ given by

$$F_1(\mathbf{q}, \mathbf{Q}) = \sum_{\alpha=1}^{n} a_\alpha \cos(Q^\alpha t + q^\alpha), \quad H = \frac{1}{t} \sum_\alpha p_\alpha q^\alpha,$$

where a_α, $\alpha = 1, \ldots, n$ are some constants. Using Eq. (8.21), we obtain the expressions for $(\mathbf{p}, \mathbf{P})$ as

$$p_\alpha = \frac{\partial F_1}{\partial q^\alpha} = -a_\alpha \sin(Q^\alpha t + q^\alpha), \quad P_\alpha = -\frac{\partial F_1}{\partial Q^\alpha} = a_\alpha t \sin(Q^\alpha t + q^\alpha). \quad (8.34)$$

Also, find the new Hamiltonian $K(\mathbf{Q}, \mathbf{P})$ from the old Hamiltonian $H(\mathbf{q}, \mathbf{p}) = \mathbf{q} \cdot \mathbf{p} = \sum_i q^i p_i$. We want to compute $(\mathbf{Q}, \mathbf{P})$ as functions of $(\mathbf{q}, \mathbf{p}, t)$. Equation (8.34) leads to

$$Q^\alpha = -\frac{1}{t} \left(\arcsin \frac{p_\alpha}{a_\alpha} + q^\alpha \right), \quad P_\alpha = -p_\alpha t. \quad (8.35)$$

One can check that $\{Q^\alpha, Q^\beta\} = 0$ since each of Q^α depends only on (p_α, q^α, t). Also, for the same reason, $\{P_\alpha, P_\beta\} = 0$. Finally, a short calculation shows that $\{Q^\alpha, P_\beta\} = \delta_\alpha^\beta$, so the transformation (Eq. (8.35)) is canonical.

The new Hamiltonian K is also computed from the old Hamiltonian H through the formula Eq. (8.22). Note that you need to compute the new Hamiltonian K in the variables $(\mathbf{Q}, \mathbf{P})$ and not the starting variables $(\mathbf{q}, \mathbf{p})$. To do that, we use Eq. (8.34) and provide the inverse mapping to (8.35):

$$q^\alpha = \arcsin \frac{P_\alpha}{a_\alpha t} - Q^\alpha t, \quad p_\alpha = -\frac{P_\alpha}{t}. \quad (8.36)$$

Additionally, we compute

$$\frac{\partial F_1}{\partial t} = -\sum_\alpha a_\alpha Q^\alpha \sin(Q^\alpha t + q^\alpha) = -\sum_\alpha \frac{P_\alpha Q^\alpha}{t}.$$

Then, the new Hamiltonian is given by

$$K(\mathbf{Q}, \mathbf{P}, t) = H + \frac{\partial F_1}{\partial t} = \frac{1}{t} \sum_\alpha p_\alpha q^\alpha - \frac{P_\alpha Q^\alpha}{t}$$

$$= \frac{1}{t} \sum_\alpha -\frac{P_\alpha}{t} \left(\arcsin \frac{P_\alpha}{a_\alpha t} - Q^\alpha t \right) - \frac{P_\alpha Q^\alpha}{t} = -\frac{1}{t^2} \sum_\alpha P_\alpha \arcsin \frac{P_\alpha}{a_\alpha t}. \quad (8.37)$$

You can see that guessing the generating functions does not usually provide a nice new Hamiltonian. In the particular case considered here, we managed to get some cancellations and obtained a new Hamiltonian that is independent of Q^α. No simplifications of the new Hamiltonian are to be expected in general. In Chap. 9, we will design a more scientific way to derive a generating function that radically simplifies the Hamiltonian by writing a certain PDE—the Hamilton-Jacobi equation.

Another use of the generating function is to find the new Hamiltonian given the transformation. If the transformation $(\mathbf{q}, \mathbf{p}) \rightarrow (\mathbf{Q}, \mathbf{P})$ is known to be canonical, one can often find the generating function F analytically and compute the new Hamiltonian. We present a basic example of this type in Problem 3 at the end of this section, where we state explicitly what type of generating function to look for. These types of problems tend to be quite complicated: given a canonical transformation, it is not always clear which type of generating function to use for a successful solution and which variables among $(\mathbf{q}, \mathbf{p}, \mathbf{Q}, \mathbf{P})$ should be taken as independent. We shall not dwell more on this problem at this point; a reader can find a wealth of examples illustrating different types of canonical transformations in [8].

8.3   Darboux' and Liouville's Theorem and Conservation of Phase Volume

First, we formulate Darboux's theorem on the local structure of Hamilton's equations through its relationship with the symplectic form. We call a two-form nondegenerate if, for any vector field Y, and given vector field X, $\omega(X, Y) = 0$, and then $X = 0$. The form is antisymmetric if $\omega(X, Y) = -\omega(Y, X)$. We also remind the reader that the form is closed if $d\omega = 0$. Darboux's theorem states the following important result.

Theorem 8.6 *Suppose a manifold M has an antisymmetric, nondegenerate, bilinear, closed two-form ω (a symplectic form). Then, this form is locally equivalent to the form $\omega = dq^\alpha \wedge dp_\alpha$ in some local coordinates.*

Darboux's theorem is actually stronger than we presented above, stating that one can choose the coordinate q^1 or p_1 arbitrarily and then build out the rest of the coordinates. The crucial meaning of Darboux's theorem is that every two-form is locally canonical, and we can thus build Hamiltonian mechanics on any manifold that can be endowed with some two-form. A reader interested in these mathematical and geometric aspects of mechanics will enjoy further reading about *symplectic geometry* [53–55].

Let us now consider the geometric meaning of the symplectic form ω. For a two-dimensional plane (q^1, p_1), the quantity $dq^1 \wedge dp_1$ represents an infinitesimal volume on the plane, and for a given domain Ω in that space, $\int_\Omega dq^1 \wedge dp_1$ is the area of that domain on the plane. To compute the volume in $2n$-dimensional phase space, we need to integrate the form

$$dq^1 \wedge dp_1 \wedge dq^2 \wedge dp_2 \wedge \ldots dq^n \wedge dp_n = (\text{const}) \times \underbrace{\omega \wedge \ldots \wedge \omega}_{n \text{ times}} = (\text{const}) \times \omega^n.$$

(8.38)

The volume in phase space is then proportional to $\int_V \omega^n$.

The first important result is the conservation of the symplectic two-form ω under the Hamiltonian flow. Indeed,

$$\dot{\omega} = d\dot{q}^\alpha \wedge dp_\alpha + dq^\alpha \wedge d\dot{p}_\alpha = d\frac{\partial H}{\partial p_\alpha} \wedge dp_\alpha - dq^\alpha \wedge d\frac{\partial H}{\partial q^\alpha}$$

$$= \frac{\partial^2 H}{\partial p_\alpha \partial p_\beta} dp_\beta \wedge dp_\alpha + \frac{\partial^2 H}{\partial p_\alpha \partial q^\beta} dq^\beta \wedge dp_\alpha \qquad (8.39)$$

$$+ \frac{\partial^2 H}{\partial q^\alpha \partial q^\beta} dq^\beta \wedge dq^\alpha + \frac{\partial^2 H}{\partial q^\alpha \partial p_\beta} dp_\beta \wedge dq^\alpha = 0.$$

Denote by $\Phi(\mathbf{q}_0, \mathbf{p}_0, t)$ the solution $(\mathbf{q}, \mathbf{p})$ starting at $(\mathbf{q}(0), \mathbf{p}(0)$ measured at time t. Then, according to Eq. (8.39), ω is conserved under the transformation $\Phi(\mathbf{q}, \mathbf{p}, t)$. Indeed, the terms involving the second derivatives with respect to (q^α, q^β) and (p_α, p_β) vanish by antisymmetry of wedge product, and the cross-derivative terms cancel. Following Theorem 8.3, the transformation Φ is canonical and thus satisfies the symplecticity condition (Eq. (8.7)). Therefore, we have proved the following lemma.

Lemma 8.7 (Poincaré (1899) Result on Symplecticity of Phase Flow) *Suppose $\Phi(\mathbf{q}, \mathbf{p}, t)$ is the phase space transformation given by the phase flow generated by some Hamiltonian at time t. Then, the transformation given by the phase flow is canonical for any t, and the gradients of Φ with respect to $\mathbf{u} = (\mathbf{q}, \mathbf{p})$ satisfy*

$$\left(\frac{\partial \Phi}{\partial \mathbf{u}}\right)^T \mathbb{J} \frac{\partial \Phi}{\partial \mathbf{u}} = \mathbb{J}, \quad \mathbb{J} := \begin{pmatrix} \mathbf{0}_{n \times n} & \mathbb{I}_{n \times n} \\ -\mathbb{I}_{n \times n} & \mathbf{0}_{n \times n} \end{pmatrix}. \qquad (8.40)$$

See also [1], Chap. 6, p. 239, and [56], Chap. VI, p. 184.

Since the form ω is conserved, any power of ω is conserved, so the phase flow conserves $\omega^2, \omega^3, \ldots, \omega^n$. As we discussed, ω^n has particular importance since it represents the phase volume. Thus, if we take the volume $V(t)$ that is evolving according to the equations of motion, $\int_{V(t)} \omega^n$ is also conserved.

Liouville's theorem states are often formulated in terms of the density of states $\rho(\mathbf{q}, \mathbf{p}, t)$. That function, multiplied by the volume element ω^n, is the number of particles in a given volume ω^n. The theorem states that the density $\rho(\mathbf{q}, \mathbf{p}, t)$ satisfies the PDE in phase space:

$$\frac{\partial \rho}{\partial t} + \frac{\partial}{\partial p_\alpha}(\rho \dot{p}_\alpha) + \frac{\partial}{\partial q^\alpha}(\rho \dot{q}^\alpha) = 0. \qquad (8.41)$$

Indeed, for any function $\rho(\mathbf{q}, \mathbf{p})\omega^n$ is the number of particles in a given volume ω^n. That number of particles, when followed along the flow, is constant since the particles are expected to neither appear nor disappear from the volume. Then, $\frac{d}{dt}(\rho\omega^n) = \dot{\rho}\omega^n = 0$ on the solutions, which gives

$$
\begin{aligned}
\frac{d\rho}{dt} &= \frac{\partial\rho}{\partial t} + \{\rho, H\} = \frac{\partial\rho}{\partial t} + \frac{\partial\rho}{\partial\mathbf{q}} \cdot \frac{\partial H}{\partial\mathbf{p}} - \frac{\partial\rho}{\partial\mathbf{p}} \cdot \frac{\partial H}{\partial\mathbf{q}} \\
&= \frac{\partial\rho}{\partial t} + \frac{\partial}{\partial q^\alpha}\left(\rho \cdot \frac{\partial H}{\partial p_\alpha}\right) - \frac{\partial}{\partial p_\alpha}\left(\rho \cdot \frac{\partial H}{\partial q^\alpha}\right) = 0,
\end{aligned}
\tag{8.42}
$$

which is exactly Eq. (8.41). The result that we have derived (Eq. (8.41)) plays a fundamental role in statistical physics.

For an n-dimensional space, we can also find other integrals by taking a two-dimensional surface S with the closed contour C forming the boundary of the surface. Then, a surface integral $\int_S \omega$ represents an integral over the surface. If we take the contour $C(t)$ to be evolving in the flow along the curves of the solutions and S evolving in such a way that the boundary of S is also evolving as $C(t)$, then the integral $\int_{S(t)} \omega = \text{const}$. There are other integrals of motion one can find that are related to the integrals of powers of ω^k on appropriate $2k$-dimensional volumes, with $1 < k < n$. These integrals play an important role in integrable systems and modern mechanics and are known as the generalizations of the Poincaré's (-Cartan's) integral invariants we have encountered in Chap. 7.

8.4 Looking Forward

Since their discovery, canonical transformations continue to play an important role in mechanics. For example, fundamental works on the stability of Hamiltonian systems [57, 58] use canonical transformations as a crucial part of considerations. There are numerous applications of canonical transformations in celestial mechanics [59]. Out of many other applications of canonical transformation, we would like to mention applications to machine learning.

Let us consider the following problem: you observe the dynamics of some kind of system, and you know that it is Hamiltonian, but you do not know the Hamiltonian itself. In other words, you have a sequence of N data pairs from observations where the initial point goes to the final point after time Δt_i, such as $(\mathbf{q}_{0,i}, \mathbf{p}_{0,i}) \rightarrow (\mathbf{q}_{f,i}, \mathbf{p}_{f,i})$, $i = 1, \ldots, N$. These data pars can be obtained, for example, from snapshots of a single trajectory or several trajectories in the phase space. The goal is to predict the dynamics of the problem, as a continuation of either a single trajectory or over the whole phase space. Such problems appear in modeling and experiments when the exact Hamiltonian is not known due to, for example, uncertainty in masses, spring constants, nonlinear elasticity laws for the springs, and other uncertainties that are always present in a realistic problem.

Several ways to solve this problem were suggested in the literature. One way is to discover the equations of motion themselves using neural networks that respect the structure of Hamiltonian systems (Hamiltonian neural networks, or HNNs) [60]. Another way to solve this problem is to derive the transformations of phase space that respect the symplectic structure as given by Eq. (8.7) [61]. Finally, a way to solve this problem using canonical transformation was derived in [62, 63]. These methods can be extended to more general Poisson systems. A solution based on reducing a Poisson system to a canonical system was presented in [64]. A learning model based on the composition of transformations that are exactly Poisson for certain classes of systems, called Lie-Poisson systems, was presented in [65, 66]. Similarly, a way to learn the dynamics using a generalization of canonical transformation for Poisson systems was done as [67, 68]. Thus, canonical transformations continue to play an important role in modern mathematics, mechanics, engineering, machine learning, and other fields.

8.5 Practice Problems

1. Consider the transformation $Q = \frac{1}{2}(q^2 + p^2)$, $P = \Phi(q, p)$. Find the form of function Φ so the transformation is canonical.
   Hint. Use polar coordinates in the (q, p) plane.
2. Consider the transformation of variables of an n-dimensional system:

$$Q_i = q_i - \gamma_i t \log(\gamma_i p_i t), \quad P_i = p_i, \qquad (8.43)$$

   where γ_i are some constants. Is that transformation canonical?
3. Consider the transformation of n-dimensional coordinates $Q_i = \arctan(p_i q_i)$, $P_i = -[1 + (p_i q_i)^2] \log q_i$.

   (a) Show that the transformation is canonical.
   (b) Find (P_i, p_i) as functions of (Q_i, q_i). Using generating functions of type 1, find the generating function $F_1 = F_1(\mathbf{q}, \mathbf{Q}, t)$.

4. Consider the transformations of type 3 with $(\mathbf{p}, \mathbf{Q})$ as independent variables. (We put all indices down for simplicity).

   (a) Find the expressions for the canonical transformation $\mathbf{q} = \mathbf{q}(\mathbf{p}, \mathbf{Q}, t)$, $\mathbf{P} = \mathbf{P}(\mathbf{p}, \mathbf{Q}, t)$ given by the generating function $F_3 = \sum_{i=1}^{n} C_i \log(t p_i) \log(t Q_i)$.
   (b) Using the solution obtained in part (a), invert the mapping and express the transformation in coordinates $\mathbf{Q} = \mathbf{Q}(\mathbf{p}, \mathbf{q}, t)$, $\mathbf{P} = \mathbf{P}(\mathbf{p}, \mathbf{q}, t)$.

Chapter 9
Hamilton-Jacobi Equation

9.1 Derivation of the Hamilton-Jacobi Equation

Our goal is to compute a canonical transformation that transforms the original system with the Hamiltonian H to the new system with the Hamiltonian K. The simplest Hamiltonian one can achieve is a constant, or, equivalently, $K = 0$. Then, in the new coordinates, $\dot{\mathbf{P}} = \mathbf{0}$ and $\dot{\mathbf{Q}} = \mathbf{0}$, so $\mathbf{P}$ and $\mathbf{Q}$ are constants. Let us consider the generating function approach and treat the case when $(\mathbf{q}, \mathbf{Q})$ are independent variables, as derived in the system (Eq. (8.21)). Remembering the new expression for the Hamiltonian (Eq. (8.22)), the equation satisfying by the generating function F is, in that case,

$$\frac{\partial F(\mathbf{q}, \mathbf{Q}, t)}{\partial t} + H(\mathbf{q}, \mathbf{p}(\mathbf{q}, \mathbf{Q}), t) = 0. \tag{9.1}$$

From Eq. (8.21), we observe that $\mathbf{p} = \frac{\partial F}{\partial \mathbf{q}}$, and we substitute these derivatives of F with respect to $\mathbf{q}$ in the Hamiltonian. Thus, the equation satisfied by the generating function F is the following partial differential equation:

$$\frac{\partial F}{\partial t} + H\left(\mathbf{q}, \frac{\partial F}{\partial \mathbf{q}}, t\right) = 0. \tag{9.2}$$

For example, consider a set of n interacting particles having the identical mass m in potential $U(\mathbf{q})$, with $\mathbf{q} = (q^1, \dots, q^n)$. The Hamiltonian H and the particular form of Eq. (9.2) for F for this system are

$$H = \frac{|\mathbf{p}|^2}{2m} + U(\mathbf{q}) \quad \Rightarrow \quad \frac{\partial F}{\partial t} + \frac{1}{2m}\left|\frac{\partial F}{\partial \mathbf{q}}\right|^2 + U(\mathbf{q}) = 0. \tag{9.3}$$

© The Author(s), under exclusive license to Springer Nature Switzerland AG 2025
V. Putkaradze, *A Concise Introduction to Classical Mechanics*,
Surveys and Tutorials in the Applied Mathematical Sciences 16,
https://doi.org/10.1007/978-3-031-84977-0_9

That is, clearly, a nonlinear equation for $F(\mathbf{q}, t)$, in general, is quite challenging to solve.

Equation (9.2) plays an important role in mechanics. If one can find a solution to this equation, one can also solve the corresponding Hamiltonian system completely. As we can see, it is sufficient to look for solution $F(\mathbf{q}, t; \mathbf{Q})$ having $\mathbf{Q}$ as a set of parameters since Eq. (9.2) does not include any derivatives concerning $\mathbf{Q}$. A particular solution of the Hamilton-Jacobi equation, leading to the canonical transformation with $K = 0$, is sometimes called *the complete integral of the Hamilton-Jacobi equations*. Finding the complete integral of the Hamilton-Jacobi equation is equivalent to finding a general solution of the canonical equations. This result is known under the name *the Jacobi theorem*.

Remark 9.1 (On Connection of the Solution of F in Eq. (9.2) with the Action S)
One can prove that the action in Hamilton's principle $S = \int_{t_0}^{t} L(\mathbf{q}, \dot{\mathbf{q}}, s)ds$ satisfies the same equation as Eq. (9.2), so $S = F+\text{const}$. Thus, the action satisfies the Hamilton-Jacobi equation, and Eq. (9.2) is often written using the variable S instead of F.

9.2 Separable Systems

Let us look at some ways to solve Eq. (9.2) or at least simplify it to make it more treatable. First, let us look at the case when $H = H(\mathbf{q}, \mathbf{p})$, so H is independent of time. Let us look for a solution for F in the form $F = T(t) + W(\mathbf{q}; \mathbf{Q})$ (semicolon means that $\mathbf{Q}$ is a parameter, not the variable to be differentiated). Then, Eq. (9.2) leads to a sum of two terms, and since they depend on different variables, t, and $\mathbf{q}$, they must be constants:

$$T'(t) + H\left(\mathbf{q}, \frac{\partial W}{\partial \mathbf{q}}\right) = 0 \Rightarrow$$

$$T'(t) = -E = \text{const} \Rightarrow \quad T = -Et, \quad H\left(\mathbf{q}, \frac{\partial W}{\partial \mathbf{q}}\right) = E. \tag{9.4}$$

It is natural to interpret the constant E as the energy since, for the case when Hamiltonian is independent on time, H is conserved.

The last equation of Eq. (9.4) is a nonlinear partial differential equation, independent of time, but involving nonlinear expression for derivatives of W. For the case of Eq. (9.3), the last equation of Eq. (9.4) becomes a version of *Eikonal equation*:

$$\frac{1}{2m}\left|\frac{\partial W}{\partial \mathbf{q}}\right|^2 + U(\mathbf{q}) = E. \tag{9.5}$$

Let us suppose that H is such that the coordinate q^μ and $p^\mu = \frac{\partial F}{\partial q^\mu}$ appear together in a combination given by a function $\psi(q^\mu, p^\mu)$. Let us call $\mathbf{q}_1$ the set of coordinates of $\mathbf{q}$ except for the coordinate q^μ. The Hamilton-Jacobi equation (Eq. (9.4)) is written as

$$H\left(\mathbf{q}_1, \frac{\partial W}{\partial \mathbf{q}_1}, \psi\left(q^\mu, \frac{\partial W_\mu}{\partial q^\mu}\right)\right) = E. \tag{9.6}$$

Then, we look for solutions in the form

$$W = W_\mu(q^\mu; \mathbf{Q}) + W_1(q^\alpha; \mathbf{Q}), \quad \alpha = 1, \ldots, n; \, \alpha \neq \mu. \tag{9.7}$$

The variable q^μ will only enter into the function ψ and nowhere else in Eq. (9.4). Let us call $\mathbf{q}_1$ the set of coordinates of $\mathbf{q}$ except for the coordinate q^μ. Therefore, that function must be constant, and so must be the rest of the Hamiltonian dependent on all other coordinates:

$$\psi\left(q^\mu, \frac{\partial W_\mu}{\partial q^\mu}\right) = A_\mu, \quad H\left(\mathbf{q}, \frac{\partial W_1}{\partial \mathbf{q}}, A_\mu\right) = E. \tag{9.8}$$

It is useful to denote $A_\mu = Q_\mu$ to make a more clear connection with canonical transformations. Note that the Eq. (9.4) in n variables separates into n single equation for q^μ (still nonlinear) and equation for $n-1$ variables given by Eq. (9.8). Normally, each of these equations can be expressed as

$$W'_\mu(q^\mu) = \Psi_m u(q^\mu; \mathbf{Q}), \quad W_\mu(q^\mu) = \int^{q^\mu} \Psi_m u(s; \mathbf{Q}) \mathrm{d}s. \tag{9.9}$$

From Eq. (9.9), the complete integral of the Hamilton-Jacobi equation can be found as

$$F = T(\mathbf{Q}, t) + \sum_{\alpha=1}^{n} W_\alpha(q_\alpha; \mathbf{Q}). \tag{9.10}$$

When the Hamiltonian is separable *and* doesn't depend on time, we can seek a solution in the above form $T = E(\mathbf{Q})t$. Completely separable Hamilton-Jacobi equations are quite rare. Still, these systems play a key role in mechanics since many real systems are close enough to integrable systems, and they yield opportunities for analytical progress. Remember that to find the general solution of the corresponding Hamiltonian, only a *particular* solution needs to be found. Finding anything like the general solution of the Hamilton-Jacobi equation is, for the purpose of this book, not necessary. However, for more general applications of Hamilton-Jacobi equations, a single solution may not be enough: see [69]; for other applications of Hamilton-Jacobi equations, one may encounter outside the mechanics-focused applications we study here.

There is a result called Stäckel's theorem, which provides the necessary and sufficient conditions for a Hamiltonian to be separable. Unfortunately, that condition is rather difficult to use. Several interesting and highly nontrivial examples of solutions for the Hamilton-Jacobi equations, as well as Stäcket's theorem, are presented in [70], Chapter 14.4.

Before we study particular examples, I invite you to appreciate the beauty of the Jacobi approach. If you find any solution of Eq. (9.1), no matter how you did it, whether it is a simple or complicated solution of Eq. (9.1), you have solved the corresponding Hamilton's equations completely! Some of the examples in the Practice problems section below would certainly be highly nontrivial to solve if they were written as Hamilton's canonical equations. And yet, the Hamilton-Jacobi approach gives an exact solution to these problems.

Note When you compute the examples below, it is also important to remember the non-degeneracy conditions for the generating functions of type 1 transformations (Eq. (8.23)). Due to these conditions, the complete integral F cannot have terms that are linear in $\mathbf{Q}$ since these terms vanish when taking the second derivative.

Example 9.1 (A Relativistic Free Particle) Let us now consider the case of the relativistic free particles to illustrate the practical use of the Hamilton-Jacobi equation to find explicit solutions to Hamilton's equation.

The Lagrangian of the relativistic free particle in $\mathbb{R}^3$ is given by the equation:

$$L = -m_0 c^2 \sqrt{1 - \frac{|\dot{\mathbf{q}}|^2}{c^2}}, \tag{9.11}$$

where m_0 is the constant (the mass of the particle at rest) and c is the speed of light (also constant). We will find the solution of the system with the Lagrangian (Eq. (9.11)) using the Hamilton-Jacobi equation:

1. The momentum is connected to velocity as

$$\mathbf{p} = \frac{\partial L}{\partial \dot{\mathbf{q}}} = \frac{m_0 \dot{\mathbf{q}}}{\sqrt{1 - \frac{|\dot{\mathbf{q}}|^2}{c^2}}} \quad \Rightarrow \quad |\dot{\mathbf{q}}|^2 = \frac{|\mathbf{p}|^2}{m_0^2 + \frac{|\mathbf{p}|^2}{c^2}} \quad \Rightarrow \quad \dot{\mathbf{q}} = \frac{\mathbf{p}}{\sqrt{m_0^2 + \frac{|\mathbf{p}|^2}{c^2}}}. \tag{9.12}$$

2. The Hamiltonian is

$$H(\mathbf{q}, \mathbf{p}) = \mathbf{p}\dot{\mathbf{q}} - L = \frac{m_0 |\dot{\mathbf{q}}|^2}{\sqrt{1 - \frac{|\dot{\mathbf{q}}|^2}{c^2}}} + m_0 c^2 \sqrt{1 - \frac{|\dot{\mathbf{q}}|^2}{c^2}} = m_0 c^2 \sqrt{1 + \frac{|\mathbf{p}|^2}{m_0^2 c^2}}. \tag{9.13}$$

3. The Hamilton-Jacobi equation is, as usual,

$$\frac{\partial F}{\partial t} + H\left(\mathbf{q}, \frac{\partial F}{\partial \mathbf{q}}, t\right) = 0, \quad \Rightarrow \quad \frac{\partial F}{\partial t} + m_0c^2\sqrt{1 + \frac{1}{m_0^2c^2}\left|\frac{\partial F}{\partial \mathbf{q}}\right|^2} = 0.$$

(9.14)

4. Since the Hamiltonian is independent of time, we can look for solutions $F = -Et + W(\mathbf{q})$, with $W(\mathbf{q})$ satisfying

$$H\left(\mathbf{q}, \frac{\partial W}{\partial \mathbf{q}}, t\right) = m_0c^2\sqrt{1 + \frac{1}{m_0^2c^2}\left|\frac{\partial W}{\partial \mathbf{q}}\right|^2} = E.$$

(9.15)

5. Equation (9.15) further admits separation of variables $W = \sum_{\alpha=1}^3 W_\alpha(q^\alpha)$. Then, for variables to separate, we need that $W_\alpha'(q^\alpha) = $const for every $\alpha = 1, 2, 3$. We will call this constant Q_α since F must depend on $(\mathbf{q}, \mathbf{Q}, t)$ where $\mathbf{Q}$ is a set of constants. Then, we get

$$W_\alpha = Q_\alpha q^\alpha \quad \Rightarrow \quad W = \mathbf{Q}\cdot\mathbf{q} \quad \Rightarrow \quad F = \mathbf{Q}\cdot\mathbf{q} - tc\sqrt{m_0^2c^2 + |\mathbf{Q}|^2},$$

(9.16)

since from Eq. (9.15), $E = c\sqrt{m_0^2c^2 + |\mathbf{Q}|^2}$. Also, it is useful to note that $\mathbf{Q}$ has the dimensions of momentum.

6. The function F defined by Eq. (9.16) is the generating function of the canonical transformation of type 1. From Eq. (8.21), we obtain

$$\mathbf{p} = \frac{\partial F}{\partial \mathbf{q}} = \mathbf{Q} = \text{const}, \quad \mathbf{P} = -\frac{\partial F}{\partial \mathbf{Q}} = -\mathbf{q} + \frac{\mathbf{Q}t}{c\sqrt{m_0^2c^2 + |\mathbf{Q}|^2}} = \text{const}.$$

(9.17)

The physical meaning of $\mathbf{Q}$ is the initial value of momentum $\mathbf{p}_0$, and the physical meaning of $\mathbf{P} = -\mathbf{q}_0$ is the negative of the initial value of the coordinate $\mathbf{q}_0$. Just to verify the correctness of our calculation, we see from Eq. (8.22) that $K = H + \frac{\partial F}{\partial t} = 0$ as expected.

7. Thus, the general solution of the equation for a relativistic particle is given by

$$\mathbf{p} = \mathbf{p}_0 = \text{const}, \quad \mathbf{q} = \mathbf{q}_0 + \frac{\mathbf{p}_0 t}{\sqrt{m_0^2c^2 + |\mathbf{p}_0|^2}}.$$

(9.18)

Example 9.2 Consider the Hamiltonian (all indices are down to avoid confusion with powers):

$$H = p_1^2 + q_1^2 - \cos \frac{p_2 + p_3}{q_2 + q_3} \,. \tag{9.19}$$

1. Since the Hamiltonian is independent of time, we are looking for a solution in the form

$$F = -Et + \sum_{j=1}^{3} W_j(q_j), \quad E = \left(W_1'(q_1)\right)^2 + q_1^2 - \cos \frac{W_2'(q_2) + W_3'(q_3)}{q_2 + q_3} \,. \tag{9.20}$$

2. At first sight, the variables in the Hamiltonian in Eq. (9.20) do not separate. However, notice that if we choose $W_2(q_2)$ and $W_3(q_3)$ in such a way that $W_2'(q_2) + W_3'(q_3)$ is proportional to $q_2 + q_3$, all the coordinate-dependent terms in the fraction in Eq. (9.20) cancel. We thus choose

$$W_2'(q_2) = Q_2 q_2 + Q_3, \quad W_3'(q_3) = Q_2 q_3 - Q_3, \tag{9.21}$$

leading to

$$W_2(q_2) = \frac{1}{2} Q_2 q_2^2 + Q_3 q_2, \quad W_3(q_3) = \frac{1}{2} Q_2 q_3^2 - Q_3 q_3 \,. \tag{9.22}$$

3. Then, the equation for $W_1'(q_1)$ is

$$W_1'(q_1)^2 + q_1^2 - \cos Q_2 = E = \text{const} \,. \tag{9.23}$$

We can choose the following equation for $W_1(q_1)$:

$$W_1'(q_1)^2 + q_1^2 = Q_1^2, \quad \Rightarrow$$
$$W_1 = \int_0^{q_1} \sqrt{Q_1^2 - q_1^2}\, dq_1 = \frac{1}{2} q_1 \sqrt{Q_1^2 - q_1^2} + \frac{Q_1^2}{2} \arcsin\left(\frac{q_1}{Q_1}\right) \,. \tag{9.24}$$

We have chosen Q_1^2 for the constant and not Q_1, so the formulas are simpler; of course, we could have chosen Q_1 instead. The fact that the quadrature in Eq. (9.24) is expressed in elementary functions is, of course, nice, but strictly speaking, not necessary; some of the practice problems at the end of this section are expressed as the quadratures.

4. From Eq. (9.20), we get

$$F = -(Q_1^2 - \cos Q_2)t + W_1(q_1; Q_1) + W_2(q_2; Q_2, Q_3) + W_3(q_3; Q_2, Q_3).$$
(9.25)

5. The equations for the coordinates are

$$p_1 = \frac{\partial F}{\partial q_1} = W_1'(q_1) = \sqrt{Q_1^2 - q_1^2}$$

$$p_2 = \frac{\partial F}{\partial q_2} = W_2'(q_2) = Q_2 q_2 + Q_3$$

$$p_3 = \frac{\partial F}{\partial q_3} = W_3'(q_3) = Q_2 q_3 - Q_3$$

$$P_1 = -\frac{\partial F}{\partial Q_1} = 2Q_1 t - Q_1 \arcsin\left(\frac{q_1}{Q_1}\right) = \text{const}$$

$$P_2 = -\frac{\partial F}{\partial Q_2} = \sin Q_2 t - \frac{1}{2}\left(q_2^2 + q_3^2\right) = \text{const}$$

$$P_3 = -\frac{\partial F}{\partial Q_3} = -q_2 + q_3 = \text{const.}$$
(9.26)

Expressions (Eq. (9.26)) represent the complete solution of Hamilton's equations arising from the Hamiltonian (Eq. (9.19)). The first three equations connect the momenta (p_1, p_2, p_3) with the coordinates (q_1, q_2, q_3). Each one of these expressions is written as some combination of variables (p_i, q_i) equal to a constant, corresponding to a conservation law. The last three expressions of Eq. (9.26) yield the connections between the coordinates $\mathbf{q}$ and time t. These expressions are implicit, connecting several coordinates to time t. In principle, these relationships could be inverted to provide explicit expressions $\mathbf{q}(t; \mathbf{Q}, \mathbf{P})$, although it is rarely possible to do it in practice.

Example 9.3 Consider a slightly modified Hamiltonian from the previous example:

$$H = \left(p_1^2 + q_1^2\right)\sin t - t\cos\frac{p_2 + p_3}{q_2 + q_3}.$$
(9.27)

1. The Hamiltonian (Eq. (9.27)) depends explicitly on time, so we take

$$F = T(t) + \sum_{j=1}^{3} W_j(q_j),$$

$$T'(t) + \sin t\left(\left(W_1'(q_1)\right)^2 + q_1^2\right) - t\cos\frac{W_2'(q_2) + W_3'(q_3)}{q_2 + q_3} = 0.$$
(9.28)

2. Proceeding as in the previous example, we compute $W_2(q_2)$ and $W_3(q_3)$ as in Eq. (9.21) and $W_1(q_1)$ as in Eq. (9.24).
3. The equation for the time dependence is

$$T'(t) + \sin t\, Q_1^2 - t \cos Q_2 = 0, \quad \Rightarrow \quad T(t) = \cos t\, Q_1^2 + \frac{t^2}{2} \cos Q_2.$$

$$(9.29)$$

4. Therefore, the solution of the Hamilton-Jacobi equation is

$$F = \cos t\, Q_1^2 + \frac{t^2}{2} \cos Q_2 + W_1(q_1; Q_1) + W_2(q_2; Q_2, Q_3) + W_3(q_3; Q_2, Q_3).$$

$$(9.30)$$

5. The solution for the coordinates is obtained from Eq. (9.30):

$$p_1 = \frac{\partial F}{\partial q_1} = W_1'(q_1) = \sqrt{Q_1^2 - q_1^2}$$

$$p_2 = \frac{\partial F}{\partial q_2} = W_2'(q_2) = Q_2 q_2 + Q_3$$

$$p_3 = \frac{\partial F}{\partial q_3} = W_3'(q_3) = Q_2 q_3 - Q_3$$

$$P_1 = -\frac{\partial F}{\partial Q_1} = -2 \cos t\, Q_1 - Q_1 \arcsin\left(\frac{q_1}{Q_1}\right) = \text{const}$$

$$P_2 = -\frac{\partial F}{\partial Q_2} = \frac{t^2}{2} \sin Q_2 - \frac{1}{2}\left(q_2^2 + q_3^2\right) = \text{const}$$

$$P_3 = -\frac{\partial F}{\partial Q_3} - -q_2 + q_3 = \text{const}$$

$$(9.31)$$

Expressions (Eq. (9.31)) present the complete solution of Hamilton's equation for the Hamiltonian (Eq. (9.27)).

9.3 Action-Angle Variables

Suppose the generating function can be written in the form Eq. (9.10). According to the equations for obtaining that momenta from generating function (Eq. (8.21)), the momenta are given by

$$p_\alpha = \frac{\partial W_\alpha(q^\alpha; \mathbf{Q})}{\partial q^\alpha} = f_\alpha(q^\alpha; \mathbf{Q}), \quad \alpha = 1, \ldots, n.$$

$$(9.32)$$

Thus, the full motion in the $2n$-dimensional space is split into the motion on individual planes (q^α, p_α), i.e., exactly n one-dimensional dynamical systems. As we have seen, on any of these planes, the form $\omega_\alpha = dq^\alpha \wedge dp_\alpha$ is the infinitesimal area. Suppose the Hamiltonian also does not depend on time. Because the motion is strictly two dimensional for every plane (q^α, p_α), there cannot be any chaotic behavior, and if a trajectory intersects itself, it must form a closed curve. Consider a closed curve trajectory C_α on that plane (q^α, p_α), and compute

$$J_\alpha = \int_{\text{int } C_\alpha} dq^\alpha \wedge dp_\alpha = \text{area surrounded by the curve } C_\alpha . \tag{9.33}$$

Since ω_α is constant along the motion, and C_α is invariant, the area of the curve is not changing, and thus, there are exactly n first integrals (constants of motion) J_α. These variables are called *actions*. Moreover, every closed trajectory on a plane is topologically equivalent to a circle, and we can introduce an angle-like variable that measures the coordinate along the circle. When action-angle variables exist, the motion occurs on the n-dimensional torus.

Let us make the canonical transformation in the plane (q^α, p_α) to the variables J_α and its canonically conjugate variable, which we will call φ_α. Then, the transformation $(q^\alpha, p_\alpha) \rightarrow (J_\alpha, \varphi_\alpha)$ is canonical, and then $H = H(J_1, \ldots, J_n)$ and is independent of φ_α. If such a transformation to the action-angle variables can be found, the equations of motion become

$$\dot{J}_\alpha = 0, \quad \dot{\varphi}_\alpha = -\frac{\partial H}{\partial J_\alpha} = \omega_\alpha = \text{const}, \quad \varphi_\alpha = \omega_\alpha t + \varphi_{0,\alpha} . \tag{9.34}$$

Thus, the variables φ are rotating about the circle with constant frequencies and are therefore justifying their name *angles*.

It is not always easy to find the action-angle variables. Fortunately, there is a fundamental result due to Liouville, which states when such separation into action-angle variables is possible. We first need to define the concept of involution.

Definition 9.2 (Involution) We call two functions $F(\mathbf{q}, \mathbf{p})$ and $G(\mathbf{q}, \mathbf{p})$ being in involution if $\{F, G\} = 0$.

We call the functions $I_\alpha(\mathbf{q}, \mathbf{p}), \alpha = 1, \ldots n$ independent if

$$\text{rank} \begin{bmatrix} \dfrac{\partial I_1}{\partial \mathbf{q}} & \cdots & \dfrac{\partial I_n}{\partial \mathbf{q}} \\ \dfrac{\partial I_1}{\partial \mathbf{p}} & \cdots & \dfrac{\partial I_n}{\partial \mathbf{p}} \end{bmatrix} = n . \tag{9.35}$$

The Liouville (or Liouville-Arnol'd) theorem states the following result.

Theorem 9.3 (Liouville-Arnol'd) *Suppose a Hamiltonian system with n degrees of freedom possesses n independent (see Eq. (9.35)) constants of motion $I_1, \ldots, I_n$*

that are in involution. Then, the system is completely integrable, i.e., there are action-angle variables $(J_\alpha, \varphi_\alpha)$, $\alpha = 1, \ldots, n$, that are obtained by a canonical transformation from $(\mathbf{q}, \mathbf{p})$ coordinates.

A *Historical Note* Arnol'd has generalized the proof of Liouville theorem for a general symplectic manifold (incidentally, in his book [1]); therefore, his name was added to the classical Liouville's theorem.

We shall also note one usually stops at finding the constants of motion satisfying the conditions of Liouville's theorem. Finding the exact expressions for action and angle is usually quite algebraically involved and is rarely possible in explicit form. Let us now give some examples of Liouville-integrable systems:

1. A particle of mass $m = 1$ is moving on a solid of revolution $x^3 = f(r)$, $r = \sqrt{(x^1)^2 + (x^2)^2}$. The phase space is two dimensional, so $n = 2$. One of the constants of motion is the Hamiltonian H. The other constant of motion is obtained from Noether's integral due to the symmetry of rotation about the x^3-axis: $N = x^1\dot{x}^2 - x^2\dot{x}^1 = x^1 p_2 - x^2 p_1$. Clearly, N and H are independent and $\{N, H\} = 0$ Thus, the system is Liouville integrable.
2. Consider the example of the bead on a freely spinning hoop, considered in Sect. 3.7. The Hamiltonian is one integral of motion, with the other one being the θ-momentum given by the second equation of Eq. (3.34). Thus, the system is Liouville integrable.

9.4 Adiabatic Invariants

Let us now consider the case when an integrable system slowly changes with time. This problem relates to the highly complex problem of perturbation of Hamiltonian systems, which is described in detail in [1], Chapter 10, and which is far beyond the scope of this book. Here, we are going to just consider one-dimensional systems.

Trajectories of a one-dimensional system on the (q, p) plane with a time-independent Hamiltonian $H(q, p)$ follow the contour lines $H(q, p) = $ const. Consider a one-dimensional system evolving about an equilibrium point where $H(q, p)$ has a minimum, for example, a pendulum performing oscillations (not necessarily small oscillations) about a stable equilibrium. Suppose now this one-dimensional Hamiltonian slowly changes with time: $H = H(q, p, \epsilon t)$, $\epsilon \ll 1$. For the example of a pendulum, one could consider the case when the length l of the pendulum slowly changes with time in a prescribed fashion: $l = l(\epsilon t)$. This change could be caused, for example, by an internal mechanism slowly changing the pendulum's length or extension or contraction of the pendulum due to the prescribed change of external temperature. Technically speaking, the system is non-autonomous, so one should consider an extended phase space (q, p, t). However, for a very slow change of the Hamiltonian, one can find the quantity that remains

Fig. 9.1 Conservation of integral invariant, which is just the area inside the closed curve of the solution $H(q, p, \tau)$ =const for fixed τ. For different values of τ, the evolution of the system selects the curves that bound the same area on the phase plane

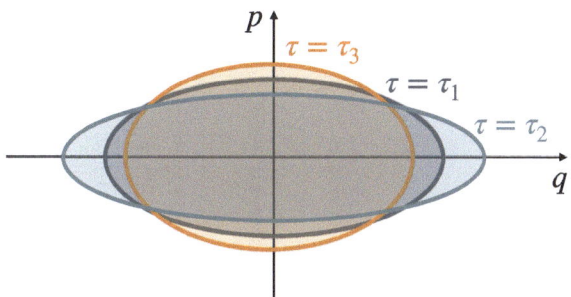

almost constant without having to consider a much more difficult problem of a fully time-dependent Hamiltonian.

We denote $\tau = \epsilon t$ and consider a closed trajectory determined by the Hamiltonian $H(q, p, \tau)$ for fixed τ. Due to the very slow change of the Hamiltonian with time, the phase portrait on the (q, p) plane will be given by $H(q, p, \tau)$ where τ is considered a parameter. As τ changes, the trajectories will slowly deform according to the change of the phase portrait, and the solution will choose one of the trajectories on the phase plane. Our goal is to find how which trajectory will be chosen at different τ by a solution. Since the Hamiltonian is time-dependent, there is no reason to expect the Hamiltonian for the solution will be conserved. However, there is a quantity, which we will call the *adiabatic invariant*, which is approximately conserved for long times. This trajectory is simply the action variable, which is defined as the area traced by the closed curve on the phase plane. One can then select the solutions at different τ based on the conservation of the adiabatic invariants, as illustrated in Fig. 9.1. Suppose for a fixed τ, the trajectory on (q, p) plane is defined as $C(\tau)$, and consider the action variable

$$I(\tau) = \int_{C(\tau)} pdq = \int_{D(\tau)=\text{int}C(\tau)} dq \wedge dp. \tag{9.36}$$

To show that I is approximately conserved, let us compute the action-angle variables. To achieve that, we find the canonical transformation from (q, p) to (I, φ). To find the generating function of this transformation, we consider the transformation to be of type 2, taking $P = I$ and q as independent variables. The generating function of the transformation $F_2(I, q)$ depends on τ as a parameter, with the variables computed as

$$p = \frac{\partial F_2}{\partial q}, \quad \varphi = -\frac{\partial F_2}{\partial I}, \tag{9.37}$$

and the new Hamiltonian K is connected with the old Hamiltonian through

$$K(I, \varphi) = H + \frac{\partial F_2}{\partial t} = H + \epsilon \frac{\partial F_2}{\partial \tau}. \tag{9.38}$$

Since (I, φ) are action-angle variables for the system with a fixed τ, the Hamiltonian H in the new variables does not depend on φ. Then, from the Hamilton's equations in (I, φ) coordinates, we get

$$\dot{I} = -\frac{\partial K}{\partial \varphi} = -\epsilon \frac{\partial^2 F_2}{\partial \varphi \partial \tau} = -\epsilon \frac{\partial}{\partial \varphi} \frac{\partial F_2}{\partial \tau}. \tag{9.39}$$

Integrating this expression for I over the period (assuming $\dot{\varphi}$ not vanishing) gives the quantity J:

$$J = \oint I \mathrm{d}\varphi, \quad \dot{J} = \epsilon \oint \frac{\partial}{\partial \varphi} \frac{\partial F_2}{\partial \tau} \mathrm{d}\varphi = 0 \tag{9.40}$$

Thus, J is of the order $O(\epsilon^2)$, and its change is of the order ϵ over the time $1/\epsilon$. Such a slowly changing quantity is called the *adiabatic invariant*.

9.5   Looking Forward

Originally, Hamilton-Jacobi equations were designed to find analytical solutions to Hamilton's equations of motion. Since then, many other applications of Hamilton-Jacobi equations have been found. Many of these applications concern numerical methods for Hamiltonian systems [71, 72]. More recently, Hamilton-Jacobi methods have been used for designing data-based integrators of Hamiltonian systems, when little information about the system is available. Instead, there are observations from which the future behavior of the system must be inferred [67, 68]. In addition to applications to mechanics, Hamilton-Jacobi equations were used in optimal control [73–76]. We also mention applications to game theory and front propagation [69, 77–79], computing propagation of fronts and level sets in numerical analysis [80] and, interestingly, finance and economics [81, 82]. As you can see, a good working knowledge of Hamilton-Jacobi equations is definitely beneficial in many fields that seem, at first sight, quite far away from mechanics.

9.6   Practice Problems

1. The Lagrangian of a particle in a magnetic field, written in spherical coordinates, is written as

$$L = \frac{m}{2} \left(\dot{r}^2 + r^2 \dot{\theta}^2 + \dot{\varphi}^2 r^2 \sin^2 \theta \right) - \dot{\varphi} \lambda \cos \theta \,,$$

where λ is a constant. Write the Hamiltonian for the system, write the Hamilton-Jacobi equation, and write the solution of this equation in quadratures using the separation of variables.

Note You don't have to compute the integrals; just write them explicitly.

2. Consider a system with the Hamiltonian

$$H = \sqrt{\sum_{\alpha} \left(p_{\alpha}^2 + \gamma_{\alpha}^2 (q^{\alpha})^2 \right)} + \varphi(t),$$

where $\gamma_{\alpha} > 0$ are some constants. Write the Hamilton-Jacobi equation and find the solution using the separation of variables. Using that solution for the Hamilton-Jacobi equation, compute the expressions for the general solution for $(\mathbf{q}(t), \mathbf{p}(t))$.

3. Consider

$$H = \frac{p_1^2}{(q^1)^2} + \frac{1}{(q^1)^2} \left(\frac{p_2^2}{(q^2)^2} + \frac{p_3^2}{(q^2 q^3)^2} \right).$$

Write the Hamilton-Jacobi equation and find the solution in quadratures using the separation of variables. Using that solution for the Hamilton-Jacobi equation, compute the general solution for $(\mathbf{q}(t), \mathbf{p}(t))$.

4. Find the action-angle variables for the harmonic oscillator:

$$H = \frac{p^2}{2m} + \frac{c}{2} q^2$$

Hint: Use the definition of the action integral in terms of the area on the phase plane.

5. Adiabatic invariants:

   (a) A particle is bouncing between two plates separated by a distance L, which is changing slowly as $L = L(\epsilon t)$, $\epsilon \ll 1$. Assuming the particle's momentum initially is p_0, find the adiabatic invariant for this system.

   (b) A pendulum of mass m is swinging in a gravitational field. The length of the pendulum is changing slowly as $l = l(\epsilon t)$, $\epsilon \ll 1$. Find the adiabatic invariant (in quadratures).

Chapter 10
Rigid Body Dynamics

10.1 Motion of a Rigid Body and Rotation Matrices

A rigid body is defined as an object that preserves relative distances between any two points of the body. This is in contrast to an elastic body, where the distances may change, creating elastic energy. For the rigid body, the potential energy of relative deformation is assumed to be so punishingly high that the relative position between any two points is fixed. Intuitively, we understand that the motion of a rigid body rotating about a fixed point can be just computed from the orientation of that rigid body. We will now formalize this intuitive concept through the formalism of rotation matrices.

Suppose a rigid body is fixed at point O, which we choose as the origin, as shown in Fig. 10.1. Let us mark a point $\mathbf{X}$ on the rigid body and observe where this point moves after time t. We mark this point as $\mathbf{x}(t, \mathbf{X})$. The key assumption of the rigid body is that all points of the rigid body move in unison, i.e.,

$$\mathbf{x}(t, \mathbf{X}) = \mathbb{A}(t)\mathbf{X}, \quad \mathbb{A}(t) \text{ is a } 3 \times 3 \text{ matrix.} \tag{10.1}$$

We are now going to use the fundamental assumption about the rigid body, that the distance between any two points remains the same under evolution as mentioned above. Take two points $\mathbf{X}$ and $\mathbf{Y}$ that are marked on the body. The coordinates of these points move to positions $\mathbf{x}(t, \mathbf{X}) = \mathbb{A}(t)\mathbf{X}$ and $\mathbf{y}(t, \mathbf{Y}) = \mathbb{A}(t)\mathbf{Y}$ in space according to Eq. (10.1). Then, remembering that $\mathbf{a} \cdot \mathbf{a} = \mathbf{a}^T \mathbf{a}$ for any vector $\mathbf{a} \in \mathbb{R}^3$, we compute

$$|\mathbf{x} - \mathbf{y}|^2 = (\mathbf{x} - \mathbf{y}) \cdot (\mathbf{x} - \mathbf{y}) = (\mathbf{X} - \mathbf{Y})\mathbb{A}^T \mathbb{A}(\mathbf{X} - \mathbf{Y}) = |\mathbf{X} - \mathbf{Y}|^2 \tag{10.2}$$

for any $\mathbf{X}$ and $\mathbf{Y}$ in the body, so $\mathbb{A}^T \mathbb{A} = \mathrm{Id}$. Such matrices are called *orthogonal*.

© The Author(s), under exclusive license to Springer Nature Switzerland AG 2025
V. Putkaradze, *A Concise Introduction to Classical Mechanics*,
Surveys and Tutorials in the Applied Mathematical Sciences 16,
https://doi.org/10.1007/978-3-031-84977-0_10

Fig. 10.1 A sketch of the rigid body dynamics. The spatial coordinate frame is (x, y, z), and the system of coordinates fixed to the body is (X, Y, Z). The point O is fixed in space. Vector **A** is a fixed vector in the body frame measuring the position of the center of mass of the body. Vector $\boldsymbol{\Gamma}$ is a unit vertical vector measured in the body frame

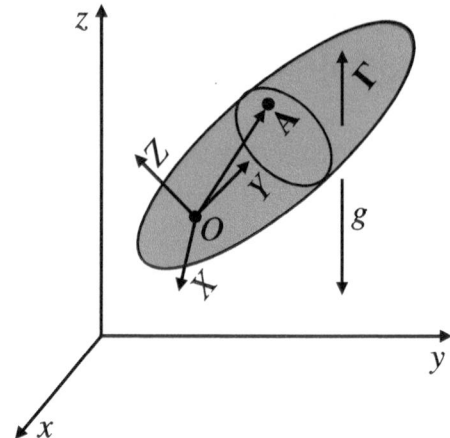

We remember that for two $n \times n$ square matrices $\mathbb{A}$ and $\mathbb{B}$, $\det(\mathbb{A}\mathbb{B}) = \det\mathbb{A} \det\mathbb{B}$, and $\det\mathbb{A}^T = \det\mathbb{A}$. Thus, for an orthogonal matrix $\mathbb{A}$,

$$\det(\mathbb{A}^T \mathbb{A}) = (\det\mathbb{A})^2 = \det \mathrm{Id} = 1,$$

so there are two options, $\det\mathbb{A} = 1$ and $\det\mathbb{A} = -1$. If we choose $\det\mathbb{A} = 1$, the matrices form a group under multiplication, which has the name of $SO(3)$—a special orthogonal group or the group of rotations. The fact that $SO(3)$ is a group means that

1. If $\mathbb{A}, \mathbb{B}$ are in $SO(3)$, then $\mathbb{A}\mathbb{B} \in SO(3)$. Indeed,

$$(\mathbb{A}\mathbb{B})^T \mathbb{A}\mathbb{B} = \mathbb{B}^T \mathbb{A}^T \mathbb{A}\mathbb{B} = \mathbb{B}^T \underbrace{(\mathbb{A}^T \mathbb{A})}_{=\mathrm{Id}} \mathbb{B} = \mathbb{B}^T \mathbb{B} = \mathrm{Id}.$$

2. There is an identity operator $\mathbb{A} = \mathrm{Id}$.
3. For any $\mathbb{A} \in SO(3)$, there is the inverse $\mathbb{A}^{-1} \in SO(3)$, which is true since $\det\mathbb{A} = 1 \neq 0$.

Let us choose fixed bases $(\mathbf{E}_1, \mathbf{E}_2, \mathbf{E}_3)$ frozen in the body, and the basis $(\mathbf{e}_1, \mathbf{e}_2, \mathbf{e}_3)$ that is stationary in the fixed frame. We define the α-th column of $\mathbb{A}$ as the coordinates of the vectors $\mathbb{E}_\alpha$ in the fixed spatial frame $(\mathbf{e}_1, \mathbf{e}_2, \mathbf{e}_3)$.

10.2 Description of Rotations of a Rigid Body Using Matrices

Let us now describe rotations using different methods. We start first with the description using rotation matrices. In the description of rotation matrices, it is useful to remember the following:

Theorem 10.1 (Euler's Theorem on Rotations) *Any matrix in $SO(3)$ can be obtained as a rotation about a fixed axis* **n** *by some angle* α.

In what follows, we shall denote rotation matrices by $R(\alpha, \mathbf{n})$. Usually, one uses the right-hand rule to determine the direction of rotations about the given axis. We can see that $\mathbb{R}(\varphi, \mathbf{n}) = -\mathbb{R}(-\varphi, -\mathbf{n})$, so the same matrix can be represented in two different ways.

In the literature, it is also common to denote the body axes $(\mathbf{E}_1, \mathbf{E}_2, \mathbf{E}_3)$ as (X, Y, Z) and denote the rotation about these axes by a given angle φ as $\mathbb{R}_X(\varphi), \mathbb{R}_Y(\varphi)$ and $\mathbb{R}_Z(\varphi)$. The rotation matrices about given coordinate axes X, Y, or Z are written in coordinates as follows:

$$\mathbb{R}_X(\varphi) = \begin{pmatrix} 1 & 0 & 0 \\ 0 & \cos\varphi & -\sin\varphi \\ 0 & \sin\varphi & \cos\varphi \end{pmatrix},$$

$$\mathbb{R}_Y(\varphi) = \begin{pmatrix} \cos\varphi & 0 & -\sin\varphi \\ 0 & 1 & 0 \\ \sin\varphi & 0 & \cos\varphi \end{pmatrix}, \tag{10.3}$$

$$\mathbb{R}_Z(\varphi) = \begin{pmatrix} \cos\varphi & -\sin\varphi & 0 \\ \sin\varphi & \cos\varphi & 0 \\ 0 & 0 & 1 \end{pmatrix}.$$

These matrices transform any given vector from its initial value $\mathbf{v}_0$ to its final value $\mathbf{v} = \mathbb{R}\mathbf{v}_0$.

The description of rotation matrices using Euler angles consists of a sequence of three rotations about the body axes. The sequence of rotations is denoted by a sequence of three letters denoting the axes of rotation on each step, for example, ZXZ.

Euler angles were popular for the description of rotations some decades ago, but they do suffer from singularities of parameterization. The use of *quaternions* is much more popular now in scientific literature, including applications to computer graphics [83, 84]. We shall not focus on this subject here; a reader who is interested in an excellent introduction to quaternions is encouraged to read [3]. One also uses the language of Lie groups to write the equations in the coordinate-free form, which is especially important for information, robotics, and control; see [85] and the extensive treatises [86, 87] on the subject. For now, you can just assume that all the rotations are described by some time-dependent matrices $\mathbb{R}(t)$, which are rotation matrices for every t, so one can say that $\mathbb{R}(t)$ is a curve in $SO(3)$.

10.3   Body and Spatial Angular Velocities

During rotations, the orientation matrix of the rigid body $\mathbb{R}(t)$ changes with time. Let us consider the matrix $\dot{\mathbb{R}}$, and notice that the following two matrices are antisymmetric $\widehat{\omega} = \dot{\mathbb{R}}\mathbb{R}^T$ and $\widehat{\Omega} = \mathbb{R}^T\dot{\mathbb{R}}$ (hats above letters denote antisymmetric matrices). This can be seen by differentiating the condition on the $SO(3)$ matrices $\mathbb{R}\mathbb{R}^T = \mathrm{Id}$:

$$\frac{d}{dt}\mathbb{R}\mathbb{R}^T = 0, \quad \dot{\mathbb{R}}\mathbb{R}^T + \mathbb{R}\dot{\mathbb{R}}^T = \dot{\mathbb{R}}\mathbb{R}^T + \left(\dot{\mathbb{R}}\mathbb{R}^T\right)^T = 0 \quad \Rightarrow \widehat{\omega} + (\widehat{\omega})^T = 0$$

(10.4)

and similarly with $\widehat{\Omega}$ by differentiating $\mathbb{R}^T\mathbb{R} = \mathrm{Id}$.

Antisymmetric Matrices and Cross-Product   Suppose $\widehat{b}$ is a 3×3 antisymmetric matrix. Because of antisymmetry, the diagonal elements are 0, and the matrix can be written as

$$\widehat{b} = \begin{pmatrix} 0 & -b^3 & b^2 \\ b^3 & 0 & -b^1 \\ -b^2 & b^1 & 0 \end{pmatrix}, \quad \text{or} \quad \widehat{b}_{ij} = -\epsilon_{ijk}b_k.$$

(10.5)

Then, one can see that for any vector $\mathbf{v} = (v^1, v^2, v^3)^T$, Eq. (10.5) gives $\widehat{b}\mathbf{v} = \mathbf{b} \times \mathbf{v}$, with the vector $\mathbf{b}$ defined as $\mathbf{b} = (b^1, b^2, b^3)^T$. Thus, 3×3 antisymmetric matrices and 3-vectors are equivalent through Eq. (10.5). We will sometimes write $\widehat{b} = \mathbf{b}\times$. The mapping from vectors $\mathbf{b} = (b^1, b^2, b^3)^T$ to antisymmetric matrices $\widehat{b}$ defined by Eq. (10.5) is called the *hat map*.

There is an inverse mapping from antisymmetric matrices to vectors. This inverse matrix is obtained by reading the off-diagonal elements in the matrix $\widehat{b}$ and inserting them in the vector elements of $\mathbf{b}$. That map is denoted by the inverted hat, and we write $b^\vee = \mathbf{b}$ for any antisymmetric matrix $\widehat{b}$.

Coming back to the rotations, we can define two useful quantities:

1. $\widehat{\omega} = \dot{\mathbb{R}}\mathbb{R}^T$ is the angular velocity of rotation expressed in the spatial frame. The corresponding vector $\boldsymbol{\omega} = \left(\dot{\mathbb{R}}\mathbb{R}^T\right)^\vee$ is the vector of angular velocities observed from the spatial frame.
2. $\widehat{\Omega} = \mathbb{R}^T\dot{\mathbb{R}}$ is the angular velocity of rotations expressed in the body frame. The corresponding vector $\boldsymbol{\Omega} = \left(\mathbb{R}^T\dot{\mathbb{R}}\right)^\vee$ is the vector of angular velocities observed from the body frame.

Notation   Here and below, we have used capital letters to denote quantities as seen in the body frame and script letters to determine the quantities seen in the spatial frame.

Let us determine the rule of transformation of vectors between body and spatial frames. Suppose A is any antisymmetric matrix and $A = A^\vee$ is its vector equivalent under the inverse hat map. Let us take any vector $\mathbf{v}$ and write $A\mathbf{v} = A \times \mathbf{v}$. If we apply any rotation matrix $\mathbb{R}$ to this statement, and remember that $\mathbb{R}(\mathbf{a} \times \mathbf{b}) = \mathbb{R}\mathbf{a} \times \mathbb{R}\mathbf{b}$, we get

$$\mathbb{R}A\mathbf{v} = \mathbb{R}(A \times \mathbf{v}) = \mathbb{R}A \times \mathbb{R}\mathbf{v}. \tag{10.6}$$

On the other hand, we can notice that $A' = \mathbb{R}A\mathbb{R}^T$ is an antisymmetric matrix, and $\mathbb{R}A\mathbf{v} = \mathbb{R}A\mathbb{R}^T\mathbb{R}\mathbf{v} = A' \times \mathbb{R}\mathbf{v}$. Remembering Eq. (10.6), we obtain the law of transformations of matrices and vectors under multiplication by $\mathbb{R}$

$$A \to \mathbb{R}A\mathbb{R}^T, \quad A = A^\vee \to \mathbb{R}A^\vee. \tag{10.7}$$

Remember the expressions of the spatial $\widehat{\omega} = \dot{\mathbb{R}}\mathbb{R}^T$ and body $\widehat{\Omega} = \mathbb{R}^T\dot{\mathbb{R}}$ angular velocities. Thus, we have the following correspondence between these matrices and vectors:

$$\widehat{\omega} = \mathbb{R}\widehat{\Omega}\mathbb{R}^T \quad \text{by definition,} \quad \omega = \mathbb{R}\Omega \quad \text{by (10.6)}. \tag{10.8}$$

Similarly, any vector can be seen in both spatial and body frames. If the vector in the spatial frame is $\mathbf{v}$, the corresponding vector seen in the body frame will be $V = \mathbb{R}^T\mathbf{v}$, and, vice versa, $\mathbf{v} = \mathbb{R}V$.

Remark 10.2 (On Common Misconception) One often hears that the angular velocity, as seen in the body frame, must be zero since the body is not rotating with respect to itself. That is incorrect. The vector of angular velocity in spatial frame ω is clearly nonzero, and the same vector, seen from the body frame, cannot be zero since it is obtained by the multiplication of a rotation matrix.

We now remember Euler's theorem on rotations (Eq. (10.1)). Rodrigues' rotation formula expresses the rotation matrix in a simple and elegant explicit formula.

Lemma 10.3 (Rodrigues' Formula) *Suppose* $\mathbf{n}$ *is a unit rotation vector and* θ *is the rotation angle about that axis. Then, the rotation matrix* $\mathbb{R}$ *is computed as*

$$\mathbb{R} = \mathrm{Id} + \sin\theta\,\widehat{n} + (1 - \cos\theta)\widehat{n}\widehat{n}, \tag{10.9}$$

where $\widehat{n}$ *is the hat map of* $\mathbf{n}$ *given by* (10.5).

Proof Left as an exercise. To prove Eq. (10.9), take any vector $\mathbf{v} \in \mathbb{R}^3$ and rotate it around the axis n by the angle θ. Finally, write the rotation as the action of the matrix $\mathbb{R}$ acting on $\mathbf{v}$.

Remark 10.4 (On Representation of Rotations Using Euler Angles) In the classical literature, one often uses a representation of rotation using Euler angles. In that representation, a rotation is described by a sequence of three rotations about some axes. For example, a popular choice is to rotate the system of coordinates around the axis z, then around the new (rotated) axis x', and then around the new axis z''. The matrix of rotation will be a product of matrix rotations about the appropriate axes; see, for example, [6], Chapter 35. However, the representation of rotations by Euler angles has inherent singularities [88] and leads to quite cumbersome equations of motion. For this reason, we will not consider the Euler angle representations here.

10.4   Euler's Equations of Motion for a Rigid Body

Suppose we have a rigid body consisting of N material particles with masses m_i and positions in the body frame $\mathbf{X}_i$. The corresponding positions in the spatial frame are $\mathbf{x}_i = \mathbb{R}\mathbf{X}_i$, with $\mathbf{X}_i$ fixed. The velocity of these particles in the spatial frame is then

$$\mathbf{v}_i = \dot{\mathbf{x}}_i = \dot{\mathbb{R}}\mathbf{X}_i = \mathbb{R}\left(\mathbb{R}^T\dot{\mathbb{R}}\mathbf{X}_i\right) = \mathbb{R}\left(\mathbf{\Omega} \times \mathbf{X}_i\right). \tag{10.10}$$

The total angular momentum of these particles in the spatial frame is computed as

$$\mathsf{l} = \sum_i m_i\mathbf{x}_i \times \mathbf{v}_i = \sum_i m_i\mathbb{R}\mathbf{X}_i \times \mathbb{R}\left(\mathbf{\Omega} \times \mathbf{X}_i\right) = \mathbb{R}\sum_i m_i\mathbf{X}_i \times \left(\mathbf{\Omega} \times \mathbf{X}_i\right). \tag{10.11}$$

Note that the angular momentum in the spatial frame depends on the orientation. However, the angular momentum in the body frame is independent of orientation and is a linear operator in the *body* angular velocity:

$$\mathbf{L} = \mathbb{R}^T\mathsf{l} = \sum_i m_i\mathbf{X}_i \times \left(\mathbf{\Omega} \times \mathbf{X}_i\right) = \mathbb{I}\mathbf{\Omega}, \quad \mathbb{I}_{\alpha\beta} = \sum_i m_i\left(|\mathbf{X}_i|^2\delta_{\alpha\beta} - X_{i,\alpha}X_{i,\beta}\right). \tag{10.12}$$

That operator on the right-hand side, multiplying $\mathbf{\Omega}$ represented by a 3×3 matrix, is called the matrix, or tensor of inertia. That matrix $\mathbb{I}$ is only constant when computed in the body frame. This matrix is symmetric, $\mathbb{I}^T = \mathbb{I}$, and positive definite. Thus, the eigenvalues of this matrix (I_1, I_2, I_3) are real and positive. We can choose the axes $\mathbf{E}_i$ to coincide with the eigenvector directions of $\mathbb{I}$. Then, the matrix $\mathbb{I}$ is diagonal in that basis:

$$\mathbb{I} = \begin{pmatrix} I_1 & 0 & 0 \\ 0 & I_2 & 0 \\ 0 & 0 & I_3 \end{pmatrix}. \tag{10.13}$$

In almost all work on the subject, the matrix $\mathbb{I}$ is assumed in the diagonal form (Eq. (10.13)), since this form can be achieved without the loss of generality. We shall assume that diagonal form of $\mathbb{I}$ as well.

Triangle Inequalities for the Inertia Matrix One can show that the coefficients $I_i, i = 1, 2, 3$ satisfy triangle inequalities:

$$I_2 + I_3 \geq I_1, \quad I_1 + I_3 \geq I_2, \quad I_1 + I_2 \geq I_3. \tag{10.14}$$

Therefore, we write the expressions for the angular momenta in the spatial **l** and in the body **L** frames:

$$\mathbf{l} = \mathbb{R}\mathbb{I}\boldsymbol{\Omega}, \quad \mathbf{L} = \mathbb{I}\boldsymbol{\Omega}. \tag{10.15}$$

Given the torque in the spatial frame **t**, the conservation of angular momentum in the spatial frame is written as

$$\frac{d\mathbf{l}}{dt} = \mathbf{t}. \tag{10.16}$$

Equation (10.16) is rather awkward to use since it involves, explicitly, the rotation matrices $\mathbb{R}$. We want to rewrite it in the body frame, so we multiply Eq. (10.16) by $\mathbb{R}^T$ and use Eq. (10.15):

$$\mathbb{R}^T \frac{d}{dt}(\mathbb{R}\mathbb{I}\boldsymbol{\Omega}) = \mathbb{I}\dot{\boldsymbol{\Omega}} + \mathbb{R}^T\dot{\mathbb{R}}\mathbb{I}\boldsymbol{\Omega} = \mathbb{I}\dot{\boldsymbol{\Omega}} + \boldsymbol{\Omega} \times \mathbb{I}\boldsymbol{\Omega} = \mathbb{R}^T\mathbf{t} = \mathbf{T}. \tag{10.17}$$

This equation deserves to be written explicitly, as it forms the foundation of all work on the theory of rotation of rigid bodies and heavy tops:

$$\mathbb{I}\dot{\boldsymbol{\Omega}} + \boldsymbol{\Omega} \times \mathbb{I}\boldsymbol{\Omega} = \mathbf{T}. \tag{10.18}$$

Note that the equation for the rigid body is written in the *body frame*. Unless you have a very good reason, there is no point in writing these equations in the spatial frame: the moment of inertia of a rigid body in the spatial frame is not constant but depends explicitly on the orientation matrix, more precisely, $\mathbb{I}_{sp} = \mathbb{R}^T\mathbb{I}\mathbb{R}$. It makes the computations very awkward. The only potential exception is the case when the matrix $\mathbb{I}$ is proportional to the identity matrix. In that case, the body is (dynamically) spherically symmetric. In that case, $\mathbb{I}_{sp} = \mathbb{I} =$const. However, this is a very particular case that is rarely encountered in practice.

10.5   Euler's Equations for the Rigid Body Motion

One of the particular cases of the rigid body motion is given when the motion is free, so there is no external torque on the body, i.e., $\mathbf{T} = \mathbf{0}$ in Eq. (10.18). In that particular case, the equations of motion reduce to Euler's equations of motion:

$$\mathbb{I}\dot{\boldsymbol{\Omega}} + \boldsymbol{\Omega} \times \mathbb{I}\boldsymbol{\Omega} = \mathbf{0} . \tag{10.19}$$

These are called Euler's equations for a rigid body.

Constants of Motion   There are two constants of motion for Eq. (10.19). One is the energy $\frac{1}{2}\boldsymbol{\Omega} \cdot \mathbb{I}\boldsymbol{\Omega}$. The conservation of energy is proven as follows:

$$\frac{d}{dt}\boldsymbol{\Omega} \cdot \mathbb{I}\boldsymbol{\Omega} = \boldsymbol{\Omega} \cdot \mathbb{I}\dot{\boldsymbol{\Omega}} + \dot{\boldsymbol{\Omega}} \cdot \mathbb{I}\boldsymbol{\Omega} = \boldsymbol{\Omega} \cdot \mathbb{I}\dot{\boldsymbol{\Omega}} + \boldsymbol{\Omega} \cdot \mathbb{I}^T \boldsymbol{\Omega} = -2\boldsymbol{\Omega} \cdot (\boldsymbol{\Omega} \times \mathbb{I}\boldsymbol{\Omega}) = 0 . \tag{10.20}$$

The second constant of motion is the absolute value of the angular momentum $|\mathbb{I}\boldsymbol{\Omega}|^2$:

$$\frac{d}{dt}|\mathbb{I}\boldsymbol{\Omega}|^2 = 2\mathbb{I}\boldsymbol{\Omega} \cdot \mathbb{I}\dot{\boldsymbol{\Omega}} = -2\mathbb{I}\boldsymbol{\Omega} \cdot (\boldsymbol{\Omega} \times \mathbb{I}\boldsymbol{\Omega}) = 0 . \tag{10.21}$$

Physically, this constant of motion can be understood as the consequence of the conservation of angular momentum in the spatial frame, $\mathbf{l} = \mathbb{R}\mathbb{I}\boldsymbol{\Omega}$. Taking $|\mathbf{l}|^2$ yields exactly $|\mathbb{I}\boldsymbol{\Omega}|^2$ and hence the result (Eq. (10.21)). From the mathematical point of view, the conservation law (Eq. (10.21)) is related to the Casimir invariant of Euler's equation written in the Hamiltonian form [3, 4].

These two constants of motion are ellipsoids in the three-dimensional space of velocities $\boldsymbol{\Omega}$. The intersection of these ellipsoids selects a closed curve in that space. Thus, all solutions of Eq. (10.19) are periodic, and the motion is integrable. It is also known that the solutions can be expressed in terms of elliptic integrals.

For future reference, it is also useful to write Eq. (10.19) explicitly in terms of the components of $\boldsymbol{\Omega} = (\Omega_1, \Omega_2, \Omega_3)$:

$$\begin{cases} I_1\dot{\Omega}_1 + (I_3 - I_2)\Omega_2\Omega_3 = 0, \\ I_2\dot{\Omega}_2 + (I_1 - I_3)\Omega_3\Omega_1 = 0, \\ I_3\dot{\Omega}_3 + (I_2 - I_1)\Omega_1\Omega_2 = 0. \end{cases} \tag{10.22}$$

10.6   Heavy Top

Let us now consider the case when the motion is affected by the gravity torque acting on the center of mass of the body. Suppose the center of mass in the body frame is at $\mathbf{X} = A$, as shown in Fig. 10.1. To compute the torque, let us first define the unit vertical vector as seen from the body frame, $\boldsymbol{\Gamma} = \mathbb{R}^T \mathbf{e}_3$. Differentiating $\boldsymbol{\Gamma}$, we find

$$\dot{\boldsymbol{\Gamma}} = \dot{\overline{\mathbb{R}^T \mathbf{e}_3}} = \dot{\mathbb{R}}^T \mathbf{e}_3 = \mathbb{R}^T \mathbb{R} \mathbb{R}^T \mathbf{e}_3 = \left(\mathbb{R}^T \dot{\mathbb{R}}\right)^T \boldsymbol{\Gamma} = \widehat{\boldsymbol{\Omega}}^T \boldsymbol{\Gamma} = -\boldsymbol{\Omega} \times \boldsymbol{\Gamma}. \qquad (10.23)$$

The force acting on the center of mass in the body frame is $-mg\boldsymbol{\Gamma}$, and the torque is $\mathbf{T} = mg\boldsymbol{\Gamma} \times A$. We could have chosen $\boldsymbol{\Gamma} = -\mathbb{R}^T \mathbf{e}_3$, with the force of gravity being $mg\boldsymbol{\Gamma}$. In that case, the sign of the torque in the equations will change; both formulations are consistent. Thus, equations of motion of a heavy rigid body in a gravitational field, also known as the *heavy top* and *Euler-Poisson* equations, are

$$\mathbb{I}\dot{\boldsymbol{\Omega}} + \boldsymbol{\Omega} \times \mathbb{I}\boldsymbol{\Omega} = mg\boldsymbol{\Gamma} \times A, \qquad \dot{\boldsymbol{\Gamma}} + \boldsymbol{\Omega} \times \boldsymbol{\Gamma} = 0. \qquad (10.24)$$

The constants of motion for the heavy top are as follows:

1. $|\boldsymbol{\Gamma}|^2 = 1$, which can be seen by multiplying the equation for $\dot{\boldsymbol{\Gamma}}$ by $\boldsymbol{\Gamma}$.
2. Energy $E = \frac{1}{2}\boldsymbol{\Omega} \cdot \mathbb{I}\boldsymbol{\Omega} + mgA \cdot \boldsymbol{\Gamma}$.
3. Projection of the angular momentum on the spatial $\mathbf{e}_3$-axis: $M = \mathbb{I}\boldsymbol{\Omega} \cdot \boldsymbol{\Gamma}$.

Indeed, the conservation of $|\boldsymbol{\Gamma}|^2$, energy E, and the projection of angular momentum M is computed as

$$\frac{1}{2}\frac{d}{dt}|\boldsymbol{\Gamma}|^2 = \boldsymbol{\Gamma} \cdot \dot{\boldsymbol{\Gamma}} = \boldsymbol{\Gamma} \cdot (-\boldsymbol{\Omega} \times \boldsymbol{\Gamma}) = 0,$$

$$\dot{E} = \boldsymbol{\Omega} \cdot \mathbb{I}\dot{\boldsymbol{\Omega}} + mgA \cdot \dot{\boldsymbol{\Gamma}} = \boldsymbol{\Omega} \cdot (-\boldsymbol{\Omega} \times \mathbb{I}\boldsymbol{\Omega} + mg\boldsymbol{\Gamma} \times A) + mgA \cdot (-\boldsymbol{\Omega} \times \boldsymbol{\Gamma})$$

$$= mg\,(A \cdot (\boldsymbol{\Omega} \times \boldsymbol{\Gamma}) - A \cdot (\boldsymbol{\Omega} \times \boldsymbol{\Gamma})) = 0,$$

$$\dot{M} = \mathbb{I}\dot{\boldsymbol{\Omega}} \cdot \boldsymbol{\Gamma} + \mathbb{I}\boldsymbol{\Omega} \cdot \dot{\boldsymbol{\Gamma}} = (-\boldsymbol{\Omega} \times \mathbb{I}\boldsymbol{\Omega} + mg\boldsymbol{\Gamma} \times A) \cdot \boldsymbol{\Gamma} + \mathbb{I}\boldsymbol{\Omega} \cdot (-\boldsymbol{\Omega} \times \boldsymbol{\Gamma})$$

$$= (-\boldsymbol{\Omega} \times \mathbb{I}\boldsymbol{\Omega}) \cdot \boldsymbol{\Gamma} - (\boldsymbol{\Omega} \times \boldsymbol{\Gamma}) \cdot \mathbb{I}\boldsymbol{\Omega} = 0,$$

$$(10.25)$$

where we used the identity for the mixed product obtained by even permutation of vectors:

$$\mathbf{a} \cdot (\mathbf{b} \times \mathbf{c}) = \mathbf{c} \cdot (\mathbf{a} \times \mathbf{b}) = \mathbf{b} \cdot (\mathbf{c} \times \mathbf{a}). \qquad (10.26)$$

These three conservation laws are not sufficient for integrability. If one could find another constant of motion, then the system would have been integrable. However, the additional constant and integrability of Euler-Poisson equations have turned out to be highly elusive. There are three (or four, depending on how you count) known cases of integrability of these equations:

1. *Euler top*: $A = 0$. The equations of motion are then Eq. (10.19); see the discussion above.

2. *Lagrange top*: $I_2 = I_3$ and $A \parallel E_3$, i.e., $A = (0, 0, a)^T$. This is the case of a symmetric top. The additional constant of motion is $\Omega_3 =$const.

3. *Kovalevskaya top*: This integrability case arises when $I_1 = I_2 = 2I_3$ and $A = (a, 0, 0)$. The Kovalevskaya integral is expressed as

$$K = \xi\xi^*, \quad \xi = (\Omega_1 + i\Omega_2)^2 - mga(\Gamma_1 + i\Gamma_2)^2, \quad \xi^* = \text{comp. conj. of } \xi .$$
$$(10.27)$$

4. *Goryachev-Chaplygin top*: For this case, we require that $I_1 = I_2 = 4I_3$ and the initial conditions satisfy $M = \mathbb{I}\Omega \cdot \Gamma = 0$. Note that this case represents a constraint on the initial conditions and not the solutions since M is conserved, so the Goryachev-Chaplygin top is not a truly integrable case of Euler-Poisson equations.

Are there any more integrable cases? The answer was not known for about 100 years, despite all efforts, until V. V. Kozlov [89] proved in 1975 that the integrability is only possible in the cases listed above; more precisely, an integral that is real analytic in a small parameter cannot exist. It is, if you think about it, an incredibly powerful result. This result was further extended by S. L. Ziglin [90], who studied how the solutions of the Euler-Poisson equation behave in complex time.

10.7   Solution for the Lagrange Top

It is quite illustrative to show the solution of the Lagrange top explicitly, as it represents an elegant application of the cyclic variables in the Euler-Lagrange equations. Let us introduce the three angles of the configuration space: φ denoting the rotation of the top about the vertical axis, θ is the tilt of the top with respect to vertical, and ψ is the rotation of the top about its own axis. The schematics of our calculation are shown in Fig. 10.2.

We can deduce that the model is integrable even without doing any calculations, just purely from the considerations of symmetry. Physically, we expect that neither kinetic nor potential energy can depend on the value of the rotation angle about the vertical φ, and the rotation of the top about its axis ψ, so $L = L(\dot{\varphi}, \dot{\psi}, \dot{\theta}, \theta)$. Then, we immediately see that the corresponding momenta in φ and ψ are conserved:

$$p_\varphi = \frac{\partial L}{\partial \dot{\varphi}} = \text{const}, \quad p_\psi = \frac{\partial L}{\partial \dot{\psi}} = \text{const}. \quad (10.28)$$

The equations (Eq. (10.28)) can be solved to give expressions for $\dot{\varphi}$ and $\dot{\psi}$ in terms of θ and $\dot{\theta}$, as well as constants p_φ and p_ψ. These expressions can be used in reducing the Lagrangian $L = L(\dot{\varphi}, \dot{\psi}, \dot{\theta}, \theta)$ to be a function of $(\dot{\theta}, \theta)$ and constants p_φ and p_ψ. The reduced system becomes one dimensional with the Lagrangian $L = L_1(\dot{\theta}, \theta; p_\varphi p_\psi)$. That one-dimensional system is basically a

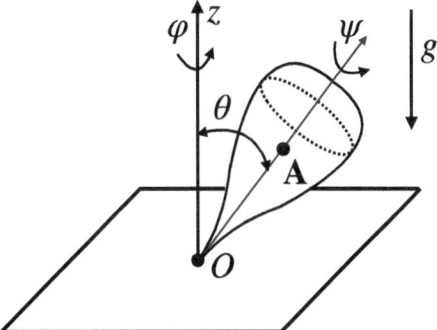

Fig. 10.2 Definition of variables used for the computation of the Lagrange top. The angles of rotation about the vertical axis z denoted as φ and about the top's own axis denoted as ψ are cyclic variables: the system is invariant with respect to rotations about these axes. The center of mass is located at $A = (a, 0, 0)$ in the body frame of reference, or the distance a from the contact point along the axis of symmetry

nonlinear oscillator. The solutions can be sketched using the phase portrait method in the $(\dot{\theta}, \theta)$ plane since the trajectories will additionally conserve the energy $E = E_1(\dot{\theta}, \theta; p_\varphi p_\psi)$ =const. Thus, the complex system of the Lagrange top will, in fact, reduce to an equation for a nonlinear oscillator, and we know this fact exclusively from symmetry without even writing down any equations.

Let us now show how to perform actual computations. A somewhat tedious calculations shows the angular velocity $\boldsymbol{\Omega} = (\mathbb{R}^T \dot{\mathbb{R}})^\vee$ is

$$\boldsymbol{\Omega} = (\dot{\varphi} \sin \theta \sin \psi + \dot{\theta} \cos \psi, \dot{\varphi} \sin \theta \cos \psi - \dot{\theta} \sin \psi, \dot{\varphi} \cos \theta + \dot{\psi})^T. \quad (10.29)$$

The potential energy is given by $V = mga \cos \theta$ for the choice of θ in Fig. 10.2. The kinetic energy is given by $T = \frac{1}{2}\boldsymbol{\Omega} \cdot \mathbb{I}\boldsymbol{\Omega}$, where $\mathbb{I} = \text{diag}(I_1, I_1, I_3)$ because of the symmetry of the Lagrange top. The Lagrangian, defined by $L = T - V$, is given by

$$L = \frac{1}{2}I_1 \left(\dot{\varphi}^2 \sin^2 \theta + \dot{\theta}^2 \right) + \frac{1}{2}I_3 \left(\dot{\varphi} \cos \theta + \dot{\psi} \right)^2 - mga \cos \theta. \quad (10.30)$$

Notice that the variables φ and ψ are cyclic, as expected, i.e., $\frac{\partial L}{\partial \varphi} = \frac{\partial L}{\partial \psi} = 0$, so the momenta with respect to these variables are conserved:

$$p_\varphi = \frac{\partial L}{\partial \dot{\varphi}} = \left(I_1 \sin^2 \theta + I_3 \cos^2 \theta \right) \dot{\varphi} + I_3 \cos \theta \dot{\psi} = \text{const}$$

$$p_\psi = \frac{\partial L}{\partial \dot{\psi}} = I_3 \left(\dot{\psi} + \dot{\varphi} \cos \theta \right) = \text{const.} \quad (10.31)$$

Equation (10.31) can be used to find $\dot\varphi$ and $\dot\psi$:

$$\dot\varphi = \frac{p_\varphi - p_\psi \cos\theta}{I_1 \sin^2\theta}, \quad \dot\psi = \frac{p_\psi}{I_3} - \frac{(p_\varphi - p_\psi \cos\theta)\cos\theta}{I_1 \sin^2\theta}. \tag{10.32}$$

We also know that the energy is conserved:

$$E = \frac{1}{2}I_1\left(\dot\varphi^2 \sin^2\theta + \dot\theta^2\right) + I_3\left(\dot\varphi \cos\theta + \dot\psi\right)^2 + mga\cos\theta = \text{const}. \tag{10.33}$$

The term multiplying I_3 in Eq. (10.33) is proportional to p_ψ and is therefore constant, not contributing to the dynamics. Substituting $\dot\varphi$ from Eq. (10.32) to Eq. (10.33), we obtain the energy reduced to $(\dot\theta, \theta)$ variables only:

$$E = \frac{p_\psi^2}{2I_3} + \frac{1}{2}I_1\dot\theta^2 + \frac{(p_\varphi - p_\psi \cos\theta)^2}{2I_1 \sin^2\theta} + mga\cos\theta = \text{const}. \tag{10.34}$$

Ignoring the constant term, this reduced energy can be viewed as an energy function for a 1D oscillator with the kinetic energy $T = \frac{1}{2}I_1\dot\theta^2$ and the potential energy

$$U(\theta) = \frac{(p_\varphi - p_\psi \cos\theta)^2}{2I_1 \sin^2\theta} + mga\cos\theta. \tag{10.35}$$

The dynamics of that system results in periodic oscillations in θ, called *nutations*. Once $\theta(t)$ is known, $\varphi(t)$ and $\psi(t)$ can be found from Eq. (10.32). Since $\dot\psi$ and $\dot\varphi$ are periodic functions in time, $\phi(t)$ and $\psi(t)$ are the sum of a periodic and linear function in time. Therefore, $\varphi(t)$ and $\psi(t)$ rotate through the whole circle, while $\theta(t)$ oscillates between the minimum and maximum values, with the values of those values depending on the initial conditions.

Sleeping Tops A sleeping top is defined as the top spinning about the vertical axis, so $\theta = 0$. It is an interesting exercise, which we will leave to the reader, to show that the sleeping top is stable only when it is spinning substantially fast about its own axis, i.e., $|\dot\psi|$ is large enough. One has to be careful in this calculation as the effective potential energy given by Eq. (10.35) exhibits a singularity at $\theta = 0$.

10.8   Looking Forward

The Euler top, or motion of the rigid body, is the most fundamental equation underpinning the rotation of satellites about their center of mass in orbit, known as the *attitude control* [91–93]. For the mathematical background of attitude control, we refer the reader to [94, 95].

From a more mathematical point of view, the example of a heavy top is a unique example of a system that is rather simple algebraically but is complex enough to test the most advanced mathematical theories that can later be applied to much more difficult problems. Thus, you will often see that papers developing modern mathematical methods for applications like fluid and plasmas treat the heavy top as one of the first examples [3, 4, 48, 96]. As it turns out, the rigid body and heavy top present interesting Hamiltonian systems—more precisely, Lie-Poisson systems. There is a particular problem on an underwater vehicle dynamics rotating about its center of mass, which is very similar to the heavy top. This problem is known as *Kichhoff's model* for underwater vehicles and models both the rotation of the vehicles and the motion of the fluid about the body as an attached mass tensor. In parallel with the heavy top, Kirchhoff's equations show rich behavior, with very few integrable cases and chaotic dynamics [97–99]. A reader can further study the theory behind this interesting problem, as well as other cases of interactions of fluids with solids, in a recent book [100].

10.9 Practice Problems

1. Rotation matrices:

   (a) Suppose $\mathbb{R}_i(\varphi_i)$ is the matrix representing the rotation of the 3D space about the axis $\mathbf{e}_i$ by angle φ_i. Compute the product of matrices $\mathbb{R}_2(\varphi_2)\mathbb{R}_1(\varphi_1)$.

   (b) Show that rotations about the same axis, performed one after another, commute. You can choose the axis to be, for example, $\mathbf{e}_1$, and compute the matrix of rotations with respect to that axis, say $\mathbb{R}_1(\varphi)$ and $\mathbb{R}_1(\theta)$. You will need to prove that $\mathbb{R}_1(\varphi)\mathbb{R}_1(\theta) = \mathbb{R}_1(\theta)\mathbb{R}_1(\varphi)$.

   (c) Suppose a rigid body with a fixed point in space also rotates about a fixed axis in space, not necessarily with a constant rate. Prove that the body and spatial angular velocities are equal at all times.

2. Explain the Dzhanibekov effect (or tennis racket theorem). Suppose a rigid body with $I_1 < I_2 < I_3$ is spun about the axis of inertia $\mathbf{E}_2$. Show that the rotations about this axis of intermediate moment of inertia are unstable. Hint: Assume $\mathbf{\Omega} = (0, \Omega_0, 0)^T + \epsilon(\alpha_1, \alpha_2, \alpha_3)^T$, with $\epsilon \ll 1$, and use Euler's equations for a rigid body. It is probably easiest to compute, e.g., $\ddot{\alpha}_3$ in terms of α_3 (correspondingly, $\ddot{\alpha}_1$ in terms of α_1), and neglect all the terms of the order ϵ^2.

   Note V. Dzhanibekov is one of the most decorated explorers of space. In June 1985, he and his fellow cosmonaut V. Savinykh docked with a dead Soviet space station, Saluyt-7, which threatened to crash to Earth, using hand controls and a laser distance meter, and brought it back to life. The space station was rotating as a rigid body, and they had to dock their spacecraft exactly at the right position and orientation with the space station without damaging the station or their own spacecraft. This is a feat that has never been attempted before or repeated since. Dzhanibekov surely knows about Euler's equations and rotations of rigid bodies!

3. Consider the Euler-Poisson equations for a heavy top (Eq. (10.24)). You can
   assume $\mathbb{I}$ is diagonal with the diagonal elements (moments of inertia) being
   $\mathbb{I} = \text{diag}(I_1, I_2, I_3)$. Suppose $F(\mathbf{\Omega}, \mathbf{\Gamma})$ is a function of the variables $\mathbf{\Omega}$ and
   $\mathbf{\Gamma}$. Compute $\frac{dF}{dt}$ on the solutions of the heavy top equations. Use this result to
   compute the rate of change for the following quantities:

   (a) $F = \mathbf{\Omega} \cdot \mathbf{\Gamma}$
   (b) $F = \mathbf{\Omega} \cdot (\mathbf{\Gamma} \times \mathbf{b})$
   (c) $F = (\mathbf{\Omega} \times \mathbf{b}) \cdot (\mathbf{\Gamma} \times \mathbf{c})$

   where $\mathbf{b}$ and $\mathbf{c}$ are fixed vectors in the body frame.

Chapter 11
Nonholonomic Constraints

11.1 Constraints and Their Validity

Consider a system with the Lagrangian $L(\mathbf{q}, \dot{\mathbf{q}}, t)$ and a set of m constraints:

$$a_\alpha^k(\mathbf{q})\dot{q}^\alpha = b^k(\mathbf{q}), \quad k = 1, \ldots, m \quad \text{sum over } \alpha. \tag{11.1}$$

The important feature of these constraints (Eq. (11.1)) is that, in general, they cannot be written as functions of coordinates only, such as $c^k(\mathbf{q}) = 0$, and thus cannot be reduced to holonomic constraints. Of course, if one differentiates a set of holonomic constraints $c^k(\mathbf{q}) = 0$ with respect to time, the result looks like Eq. (11.1) with $a_\alpha^k = \frac{\partial c^k}{\partial q^\alpha}$. However, these constraints are still holonomic. We are interested in the case when the constraints (Eq. (11.1)) are not of holonomic type.

Most often, mechanics books call any mechanical system with a set of constraints (Eq. (11.1)) such that it is not reducible to holonomic constraints, a *non-holonomic system*. We shall caution a reader about the broad application of this name to all constraints written in the form Eq. (11.1). In fact, it was shown by A. S. Karapetyan [101] and V. V. Kozlov [102] that it is not sufficient to specify a Lagrangian L and constraints (Eq. (11.1)) to write equations of motion. The *constraint realization argument* must also be specified to be able to close the system. You can read more about the realization of constraints in [24] where it is shown how formally the same-looking constraints can lead to different equations of motion.

We consider a particular case of non-holonomic constraints, appearing when there is a large viscous force pushing the system toward the constraint whenever there is a small deviation from the constraint (Eq. (11.1)). The consideration presented in this section (namely, Lagrange-d'Alembert's formula) will work in the limit of the friction force being infinitely large. However, you should always consider the limiting process by which the constraint has arisen, as other methods of treating constraints may be necessary. We will consider cases when the constraint

© The Author(s), under exclusive license to Springer Nature Switzerland AG 2025
V. Putkaradze, *A Concise Introduction to Classical Mechanics*,
Surveys and Tutorials in the Applied Mathematical Sciences 16,
https://doi.org/10.1007/978-3-031-84977-0_11

comes from either rolling without slipping or a blade moving on the ice, for which the method outlined in this section is valid. In all the cases considered here, the equations of motion can be confirmed by direct application of Newton's law (which we will not do here).

The comprehensive reference to the subject of this chapter is the book by A. M. Bloch [103], where you will find a much more detailed explanation of the mathematical principles developed here, many examples, and extensive applications to control.

11.2   Lagrange-d'Alembert's Principle

Consider a set of m constraints defined by Eq. (11.1), and the displacement generated by a system is δq^α. Suppose the force applied by the constraint k is F_α^k, and there are m constraints in total. The constraint is called *ideal* if the virtual work done by the constraint vanishes:

$$\sum_{\alpha=1}^n F_\alpha^k \delta q^\alpha = 0 \text{ for all } \alpha, \quad k = 1, \ldots, m, \tag{11.2}$$

where n is the dimension of the system. Let us enforce these conditions on variations (Eq. (11.2)) using m Lagrange multipliers λ_k. The action principle incorporating external forces gives

$$\delta \int L(\mathbf{q}, \dot{\mathbf{q}}, t)dt = -\int \sum_{k=1}^m \sum_{\alpha=1}^n \lambda_k F_\alpha^k \delta q^\alpha dt \quad \Rightarrow \quad \frac{d}{dt}\frac{\partial L}{\partial \dot{q}^\alpha} - \frac{\partial L}{\partial q^\alpha} = \sum_{k=1}^m \lambda_k F_\alpha^k. \tag{11.3}$$

The key question is how to find the forces F_α^k from Eq. (11.3). The answer is provided by
Lagrange-d'Alembert's principle: for constraints (Eq. (11.1)), and appropriate physical process leading to the constraint (strong forces of viscous friction away from the constraint), the variations satisfy

$$a_\alpha^k(\mathbf{q})\delta q^\alpha = 0, \quad \text{("dots-to-deltas")}. \tag{11.4}$$

Using the constraints (Eq. (11.4)) in the variational principle, we get

$$\delta \int L(\mathbf{q}, \dot{\mathbf{q}}, t)dt + \int \lambda_k^\alpha a(\mathbf{q})_\alpha^k \delta q^\alpha dt = 0, \quad \Rightarrow \quad \frac{d}{dt}\frac{\partial L}{\partial \dot{q}^\alpha} - \frac{\partial L}{\partial q^\alpha} = \lambda_k a(\mathbf{q})_\alpha^k. \tag{11.5}$$

Comparing Eqs. (11.3) and (11.5), we see that the reaction force by k-th constraint on the coordinate q^α is $a(\mathbf{q})^k_\alpha$. Equations (Eq. (11.5)) need to be combined with conditions (Eq. (11.1)) to compute the Lagrange multipliers λ_k. These are the non-holonomic equations of motion, or, in other words, equations of motion in the framework of Lagrange-d'Alembert's principle applied to constraints (Eq. (11.1)).

Energy Conservation for Non-holonomic Systems Let us consider the equations obtained by the variational principle (Eq. (11.5)). Consider the energy defined by Eq. (2.13), and compute the rate of change for this energy, assuming that the Lagrangian L doesn't depend on time, i.e., $\frac{\partial L}{\partial t} = 0$:

$$\dot{E} = \frac{d}{dt}\left(\dot{q}^\alpha \frac{\partial L}{\partial \dot{q}^\alpha} - L\right) = \lambda_k a^k_\alpha(\mathbf{q})\dot{q}^\alpha = \lambda_k b^k(\mathbf{q}). \tag{11.6}$$

Thus, the energy is conserved when the constraint (Eq. (11.1)) is homogeneous, i.e., whenever $b^k(\mathbf{q}) = 0$.

11.3 Examples of Systems with Non-holonomic Constraints

Rolling Vertical Disk Suppose a coin rolls on the surface without any vertical tilt, as shown in the left panel of Fig. 11.1. There are two angles of rotation, the angle of rotation about the vertical axis φ and rotation about the disk's axis θ, and the corresponding moments of inertia being I and J. The position of the center of mass on the plane is (x, y). The Lagrangian is equal to the kinetic energy:

$$L = T = \frac{1}{2}m(\dot{x}^2 + \dot{y}^2) + \frac{1}{2}I\dot{\theta}^2 + \frac{1}{2}J\dot{\varphi}^2. \tag{11.7}$$

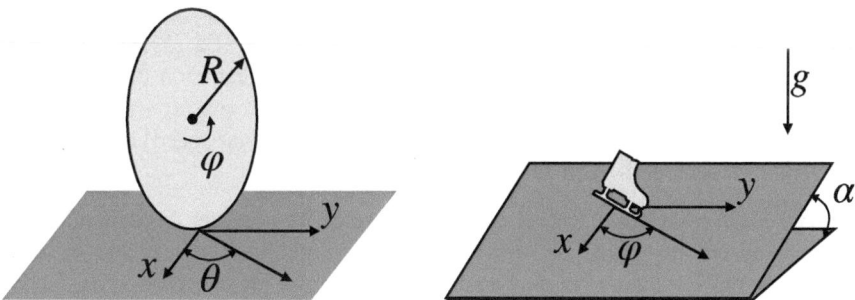

Fig. 11.1 Left panel: a vertically rolling disk on a horizontal plane. Right panel: a skate on an inclined plane

The disk rolls along the line of contact without slipping:

$$\begin{cases} \dot{x} - R\cos\varphi\dot{\theta} = 0, \\ \dot{y} - R\sin\varphi\dot{\theta} = 0. \end{cases} \tag{11.8}$$

Equations of motion are obtained using $\delta \int L\,dt = 0$ on variations satisfying

$$\begin{cases} \delta x - R\cos\varphi\delta\theta = 0, \\ \delta y - R\sin\varphi\delta\theta = 0. \end{cases} \tag{11.9}$$

We proceed as follows. Since L only depends on $(\dot{x}, \dot{y}, \dot{\varphi}, \dot{\theta})$,

$$\begin{aligned} 0 &= \int \frac{\partial L}{\partial \dot{x}}\delta\dot{x} + \frac{\partial L}{\partial \dot{y}}\delta\dot{y} + \frac{\partial L}{\partial \dot{\varphi}}\delta\dot{\varphi} + \frac{\partial L}{\partial \dot{\theta}}\delta\dot{\theta}\,dt \\ &= \int -\frac{d}{dt}\frac{\partial L}{\partial \dot{x}}\delta x - \frac{d}{dt}\frac{\partial L}{\partial \dot{y}}\delta y - \frac{d}{dt}\frac{\partial L}{\partial \dot{\varphi}}\delta\varphi - \frac{d}{dt}\frac{\partial L}{\partial \dot{\theta}}\delta\theta\,dt \\ &= \int -m\ddot{x}\delta x - m\ddot{y}\delta y - J\ddot{\varphi}\delta\varphi - I\ddot{\theta}\delta\theta\,dt \\ &= \int -m\ddot{x}R\cos\varphi\delta\theta - m\ddot{y}R\sin\varphi\delta\theta - J\ddot{\varphi}\delta\varphi - I\ddot{\theta}\delta\theta\,dt, \end{aligned} \tag{11.10}$$

where in the last line, we have substituted δx and δy from Eq. (11.9). We now collect the terms proportional to $\delta\varphi$ and $\delta\theta$ to obtain

$$\begin{aligned} \delta\varphi: \quad & J\ddot{\varphi} = 0, \\ \delta\theta: \quad & m(\ddot{x}\cos\varphi + \ddot{y}\sin\varphi) + I\ddot{\theta} = 0, \end{aligned} \tag{11.11}$$

as well as constraints (Eq. (11.8)). In principle, we are done, but this particular problem can be solved further. From the first equation, we derive that $\varphi = \omega_\varphi t + \varphi_0$. We can find $\ddot{x}$ and $\ddot{y}$ by differentiating the constraints (Eq. (11.8)), and then substitute these expressions for $\ddot{x}$ and $\ddot{y}$ into the second equation of Eq. (11.11). The resulting equation is simply $(I + mR^2)\ddot{\theta} = 0$, so $\theta = \omega_\theta t + \theta_0$. Thus, the disk rolls with constant angular velocity (and hence, constant speed) and rotates at a constant rate about the vertical axis.

Knife's Edge on a Plane (Skate on Inclined Ice) We consider the dynamics of a skate on an inclined plane, as shown in the right panel of Fig. 11.1. The variables describing the system are position and orientation on the plane, so the configuration manifold is $\mathbb{R}^2 \times S^1$. We choose the coordinates (x, y) for the position and φ for the angle of rotation. We assume that we measure the angle of the blade with the given axis. Suppose the mass of the blade is m, the moment of inertia for rotation about the axis is J, and the inclination angle of the plane is α. The Lagrangian is

$$L = \frac{1}{2}m\left(\dot{x}^2 + \dot{y}^2\right) + \frac{1}{2}J\dot{\varphi}^2 + mgy\sin\alpha. \tag{11.12}$$

The direction of the skate is $\boldsymbol{\tau} = (\cos\varphi, \sin\varphi)$, and normal to the blade is $\mathbf{n} = (-\sin\varphi, \cos\varphi)$. The velocity's $\mathbf{v} = (\dot{x}, \dot{y})^T$ projection to the direction normal to the blade must vanish, so the non-holonomic constraint $\boldsymbol{\tau} \cdot \mathbf{v}$ gives

$$\dot{x}\sin\varphi - \dot{y}\cos\varphi = 0. \tag{11.13}$$

According to Lagrange-d'Alembert's principle, the variations satisfy

$$\delta x \sin\varphi - \delta y \cos\varphi = 0. \tag{11.14}$$

Lagrange-d'Alembert's principle gives

$$\delta\int L dt = \int -\frac{d}{dt}\frac{\partial L}{\partial\dot{x}}\delta x + \left(-\frac{d}{dt}\frac{\partial L}{\partial\dot{y}} + \frac{d}{dt}\frac{\partial L}{\partial y}\right)\delta y - \frac{d}{dt}\frac{\partial L}{\partial\dot{\varphi}}\delta\varphi. \tag{11.15}$$

We now use the constraint (Eq. (11.14)) with the Lagrange multiplier λ to obtain the equations of motion:

$$m\ddot{x} = \lambda\sin\varphi, \quad m\ddot{y} = -\lambda\cos\varphi + mg\sin\alpha, \quad J\ddot{\varphi} = 0. \tag{11.16}$$

The last equation of Eq. (11.16) integrates to give $\varphi = \omega_\varphi t + \varphi_0$. We can compute λ from the first two equations.

We need to differentiate the constraint (Eq. (11.13)) and use the fact that $\dot{\varphi} = \omega_\varphi =$ const to obtain

$$\ddot{x}\sin\varphi - \ddot{y}\cos\varphi + \omega_\varphi\left(\dot{x}\cos\varphi + \dot{y}\sin\varphi\right) = 0. \tag{11.17}$$

Multiplying the first equation of Eq. (11.16) by $\sin\varphi$, the second equation by $\cos\varphi$ and subtracting, we use Eq. (11.17) to obtain

$$\lambda = -\omega_\varphi\left(\dot{x}\cos\varphi + \dot{y}\sin\varphi\right) + mg\sin\alpha\cos\varphi. \tag{11.18}$$

Normally, this would be as far as one would get in these kinds of problems; however, we can proceed further using the conservation of energy since the constraint is homogeneous, i.e., $b = 0$ in Eq. (11.1). The energy is

$$E = \frac{1}{2}m(\dot{x}^2 + \dot{y}^2) + \frac{1}{2}J\dot{\varphi}^2 - mgy\sin\alpha. \tag{11.19}$$

We use the constraint to express $\dot{x} = \dot{y} \cot \varphi$, and use the solution for $\varphi = \omega_\varphi t + \varphi_0$ to write the energy as a single ODE for $y(t)$:

$$m\dot{y}^2 \frac{1}{\cos^2(\omega_\varphi t + \varphi_0)} - mgy \sin \alpha = E - \frac{1}{J}\omega_\varphi^2 = \text{const}. \tag{11.20}$$

The solution of Eq. (11.20) can be found in terms of elementary functions (prove it!).

Chaplygin's Sleigh Chaplygin's sleigh is a model of a two-dimensional skater on horizontal ice. It consists of a two-dimensional rigid body moving along a given direction at the point of contact C and the center of mass positioned at a given distance in the direction of the skate, usually taken to be behind the skate, as illustrated in the left panel of Fig. 11.2.

The coordinates are the position of the point of contact of the blade (x, y) and rotation angle θ, so the configuration space is $\mathbb{R}^2 \times S^1$, like in the previous case. The center of mass is at $(x_c, y_c) = (x + a \cos\theta, y + a \sin\theta)$. The constraint is given by the fact that the motion on the plane occurs along the blade, just as in Eq. (11.13) *i.e.*,

$$\dot{x} \sin \theta = \dot{y} \cos \theta. \tag{11.21}$$

Again, Lagrange-d'Alembert's principle requires that the variations satisfy

$$\delta x \sin \theta - \delta y \cos \theta = 0. \tag{11.22}$$

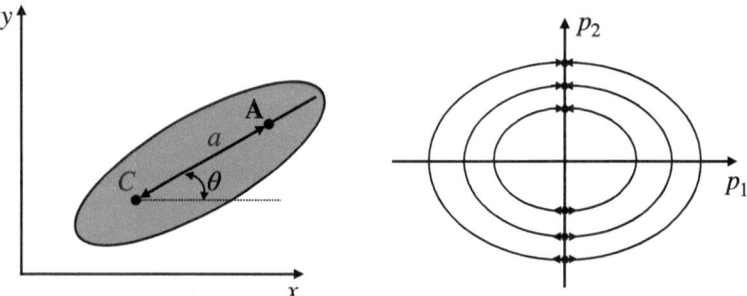

Fig. 11.2 Left panel: Chaplygin's sleigh, representing a rigid body that can move on a two-dimensional plane along a direction specified by an edge positioned at the point of contact C. The center of mass is at the distance a behind the contact point, at the direction of the skate. Right panel: a sketch of solutions of the reduced equations (Eq. (11.26)). The whole p_2-axis is the set of equilibrium points; the trajectories in (p_1, p_2) space are half-ellipses tending to $(p_1, p_2) \to (0, -p_*)$ as $t \to -\infty$ and $(p_1, p_2) \to (0, p_*)$ as $t \to \infty$

The Lagrangian is given by

$$L = \frac{1}{2}m(\dot{x}_c^2 + \dot{y}_c^2) + \frac{1}{2}J\dot{\theta}^2 = \frac{1}{2}m\left(\dot{x}^2 + \dot{y}^2 + 2a\dot{\theta}\left(\boxed{-\dot{x}\sin\theta + \dot{y}\cos\theta}\right) + a^2\dot{\theta}^2\right)$$
$$+ \frac{1}{2}J\dot{\theta}^2. \tag{11.23}$$

Notice that the expression in the box is exactly the constraint. However, we are *not* allowed to substitute the constraint in the Lagrangian and drop that term; we need to write Lagrange-d'Alembert's principle first and only substitute the constraints after the equations are derived.

The equations of motion are derived similarly to the previous example, which we leave as an exercise. Using the Lagrange multiplier λ to enforce the constraint on variations (Eq. (11.22)), we obtain

$$\delta x : m\ddot{x} - \frac{d}{dt}\left(a\dot{\theta}\sin\theta\right) = \lambda\sin\theta$$

$$\delta y : m\ddot{x} + \frac{d}{dt}\left(a\dot{\theta}\cos\theta\right) = -\lambda\cos\theta$$

$$\delta\theta : (ma^2 + J)\ddot{\theta} + ma\frac{d}{dt}\left(\boxed{-\dot{x}\sin\theta + \dot{y}\cos\theta}\right) + ma\dot{\theta}(\dot{x}\cos\theta + \dot{y}\sin\theta) = 0.$$
$$\tag{11.24}$$

Since we have performed all the differentiation, we are allowed to substitute the constraint (Eq. (11.21)) and eliminate the boxed term in the last equation of the system (Eq. (11.24)). Surprisingly, an analytical solution of these complicated-looking equations can be found. Let us define the variables:

$$p_1 = (ma^2 + J)\dot{\theta}, \quad p_2 = m(\dot{x}\cos\theta + \dot{y}\sin\theta). \tag{11.25}$$

Physically, p_1 is the angular momentum of rotation about the center of mass, and p_2 is the momentum of the blade on ice in the direction tangent to the blade. Differentiating p_2 and performing some cancellations, we obtain $\dot{p}_2 = ma\dot{\theta}^2$. Combining this result with the third equation of Eq. (11.24), we obtain equations for $p_{1,2}$ [103, 104]:

$$\dot{p}_1 = -\frac{a}{J + ma^2}p_1 p_2, \quad \dot{p}_2 = \frac{ma}{(J + ma^2)^2}p_1^2. \tag{11.26}$$

We can write the "energy-like" conserved quantity:

$$E = \frac{1}{2}\left(\frac{p_1^2}{J + ma^2} + \frac{p_2^2}{m}\right) = \text{const}. \tag{11.27}$$

Thus, the trajectories are parts of an ellipse. One can also see any point of the vertical axis $p_1 = 0$ is a critical point of Eq. (11.26). The trajectories on (p_1, p_2) plane start from the bottom half of the vertical axis $p_1 = 0$ and go to the mirror image of that point on the upper half of p_2 semiaxis using an elliptical trajectory in (p_1, p_2) plane.

11.4   Looking Forward

We have just touched briefly upon the topic of nonholonomic mechanics. The ultimate reference on the subject, as we mentioned above, is [103]; see also [105, 106]. These references contain the exposition of nonholonomic mechanics in terms of modern methods of geometry. Because of the complexity of the field, integrable (*i.e.*, completely solvable) examples of nonholonomic systems are rare [107], and thus every new integrable example is interesting. In addition to several classical examples known for some time, an amusing example of a model related to skating has been recently developed [108] and further studied in [109].

Recently, there has been a renewed interest in the theory of nonholonomic mechanics from the point of view of Hamel's method [110, 111]. This method keeps the coordinates but chooses the new "quasivelocities," which are combinations of old velocities with the transformation matrices depending on the coordinates. Such transformation potentially avoids the use of Lagrange multipliers altogether [112–114].

Another interesting direction of research is the construction of modern numerical methods preserving as much structure of nonholonomic systems as possible, which is a highly nontrivial task; see, for example, [115–117]. Further applications of the field concern the control of noholonomic systems like wheeled robots; for the detailed reference, see [103, 118]. In particular, steering Chaplygin's sleigh using a moving mass somewhat reminds the motion of a skater tracing trajectories on ice [119, 120]. Thus, nonholonomic mechanics is an actively developing, technically challenging field that holds a lot of promise for potential applications and further progress of mathematical theory inspired by this class of problems.

11.5   Practice Problems

1. Consider the (artificial) non-holonomic particle moving in three-dimensional space, with the free particle Lagrangian:

$$L = \frac{1}{2}\left(\dot{x}^2 + \dot{y}^2 + \dot{z}^2\right) \tag{11.28}$$

and non-holonomic constraint $\dot{x} = y\dot{z}$:

(a) Derive equations of motions for such a particle using Lagrange-d'Alembert's principle.

(b) Substitute naively the constraint $\dot{x} = y\dot{z}$ in the Lagrangian (Eq. (11.28)) and obtain the new "Lagrangian" without the constraints. Derive the Euler-Lagrange equation for that new Lagrangian. This method is known as the *Vakonomic* approach. Does it correspond to the result you obtained by Lagrange-d'Alembert's principle?

(c) Notice that both the Lagrangian and the constraint are invariant with respect to the coordinate shift $x \to x + a$, $z \to z + c$. From the naive application of Noether's theorem, you would expect that the momenta in the x and z direction, i.e., p_x and p_y, should be conserved, and momentum in direction y does not have to be. Is it true? (Actually, the very opposite of this statement happens to be true, which shows that it is not advisable to reach hasty conclusions for non-holonomic systems!)

(d) Compute the rate change for the energy E, which in this case happens to coincide with the Lagrangian L from Eq. (11.28), according to the equations of motion you have derived.

(e) Find the analytical formula for $y(t)$, and either $x(t)$ or $z(t)$. If you find either $x(t)$ or $z(t)$, another formula will be found in quadratures using the constraint (and, in fact, can be found explicitly, although the integrals may become a bit tedious). Hint: One way (but by no means the only way) to solve the last problem is to notice that the equation of motion connecting $x(t)$ or $z(t)$ can be written as a full derivative.

2. *Suslov's problem*: Consider a rigid body with the constraint that projection of the angular velocity on a fixed vector in the body frame **a** vanishes, i.e., $\boldsymbol{\Omega} \cdot \mathbf{a}$. This is a somewhat artificial problem, which is of great importance for pedagogical description of non-holonomic systems, but of limited practical value. The equations of motion for Suslov's problem are

$$\mathbb{I}\dot{\boldsymbol{\Omega}} + \boldsymbol{\Omega} \times \mathbb{I}\boldsymbol{\Omega} = \lambda\mathbf{a}, \quad \boldsymbol{\Omega} \cdot \mathbf{a} = 0. \tag{11.29}$$

Here, λ is the Lagrange multiplier enforcing the constraint:

(a) Compute λ explicitly in terms of $\mathbb{I}$, **a**, and $\boldsymbol{\Omega}$.
   Hint: Differentiate the constraint and use the fact that $\mathbb{I}$ is invertible and symmetric, so $\mathbf{a} \cdot \mathbf{b} = (\mathbb{I}^{-1}\mathbf{a}) \cdot (\mathbb{I}\mathbf{b})$ for any vectors **a** and **b**.

(b) Show that the energy $E = \frac{1}{2}\boldsymbol{\Omega} \cdot \mathbb{I}\boldsymbol{\Omega}$ is conserved.

(c) Suppose that $\mathbb{I}$ is diagonal: $\mathbb{I} = \text{diag}(I_1, I_2, I_3)$. Consider **a** to be along one of the inertia axes, for example, $\mathbf{a} = \mathbf{E}_1$. Show that in that case, the constraint and equations of motion give particularly simple solutions. Solve for $\boldsymbol{\Omega}(t)$ and the Lagrange multiplier λ.
   Hint. Write out the equations of motion (Eq. (11.29)) explicitly in coordinates to solve this problem.

Chapter 12
Euler-Poincaré Variational Theory for a Rigid Body

This Section May Be Skipped During the First Reading.

12.1 Symmetry of Mechanical Systems

In these notes, we have extensively studied Euler-Lagrange equations (Eq. (2.11)). The theory behind these equations is elegant and widely used but not always practical. Let us illustrate some of the difficulties in the example of the rigid body moving about its fixed center of mass in space, as we discussed in Chap. 10. The configuration manifold Q of a rigid body is the group $SO(3)$ of rotation matrices. A Lagrangian, depending on the coordinates and velocities, can be constructed and has the form $L(\mathbb{R}, \dot{\mathbb{R}})$. A naive application of the Euler-Lagrange equations with constraints will lead to the Euler-Lagrange equations for nine matrix coordinates of $\mathbb{R}$, coupled with six constraints coming from $\mathbb{R}\mathbb{R}^T = \mathrm{Id}_{3\times 3}$. While the total number of equations is three, as expected, the equations of motions obtained by this method are excessively complex. One can parameterize the group $SO(3)$ using, for example, three Euler angles, in which case the Euler-Lagrange equations will give highly nonintuitive equations for these angles. These equations will look nothing like the elegant Euler's equations for a rigid body (Eq. (10.19)). To understand how to generalize this approach by Euler, in 1901, Poincaré [121] carried out a modern derivation of these equations. That derivation is now known as the *Euler-Poincaré theory*, and we will briefly outline it here. We will derive these equations for the $SO(3)$ group, but the approach can be extended to an arbitrary Lie group.

The key to Poincaré's method is to notice that the Lagrangian is invariant with respect to arbitrary rotations of the space. More precisely, for any fixed rotation matrix $\mathbb{A} \in SO(3)$, we have $L(\mathbb{A}\mathbb{R}, \mathbb{A}\dot{\mathbb{R}}) = L(\mathbb{R}, \dot{\mathbb{R}})$. The fact that $\mathbb{R}$ is multiplied from the left by $\mathbb{R}$ comes from physics; as a rule, the dynamics of elastic and rigid bodies is left invariant. Then, the Lagrangian can be brought to a form that depends

V. Putkaradze, *A Concise Introduction to Classical Mechanics*,
Surveys and Tutorials in the Applied Mathematical Sciences 16,
https://doi.org/10.1007/978-3-031-84977-0_12

on the single variable $\widehat{\Omega} = \mathbb{R}^T \dot{\mathbb{R}}$, called the angular velocity *in the body frame*. Poincaré's method works, in fact, for any Lie group, not necessarily $SO(3)$, and has been useful for deriving the equations for complex systems consisting of interacting parts, fluids, liquid crystals, etc. The principle remains the same; only the group and configuration manifolds change. Here, we shall only consider the case when the configuration manifold Q is itself the symmetry group, as is the case for the motion of a rigid body (the case of a complete symmetry reduction). A reader who wants to go beyond a brief derivation presented here should consult [3] or [4].

12.2   Notations and Definitions

We have seen in Sect. 10.3 that the angular velocity $\widehat{\Omega} = \mathbb{R}^T \dot{\mathbb{R}}$ is an antisymmetric matrix. As we have seen in Chap. 10.3, these matrices are equivalent to vectors in three-dimensional space through the hat map. We called the mapping from vectors to antisymmetric matrices the *hat map*, with the notation $\widehat{\boldsymbol{\Omega}} = \widehat{\Omega}$. The inverse procedure, taking an antisymmetric matrix and producing a vector, is denoted as *inverse hat map* and is denoted as $\Omega^\vee = \boldsymbol{\Omega}$. In coordinates, we have $\Omega_{ij} = -\epsilon_{ijk}\Omega_k$ where ϵ_{ijk} is the completely antisymmetric tensor with $\epsilon_{123} = 1$. A useful property of the hat map relates the commutator of matrices a and b to the cross-product of vectors $\mathbf{a} = a^\vee$ and $\mathbf{b} = b^\vee$ as

$$(ab - ba)^\vee = [a, b]^\vee = \mathbf{a} \times \mathbf{b} \quad \Leftrightarrow \quad ab - ba = [a, b] = \widehat{\mathbf{a} \times \mathbf{b}}. \quad (12.1)$$

Thus, we can treat the angular velocity $\widehat{\Omega}$ to be both an antisymmetric matrix when it is defined as $\widehat{\Omega} = \mathbb{R}^T \dot{\mathbb{R}}$ and, at the same time, a three-vector using $\boldsymbol{\Omega} = \Omega^\vee = (\mathbb{R}^T \dot{\mathbb{R}})^\vee$ through the hat map. These representations are completely equivalent and are fundamental for our further discussions. The cross-product is only valid in three dimensions; for other Lie groups, there is no ready analog of a hat map, and *adjoint* and *co-adjoint* representations need to be used [4].

In addition, it is also useful to review the concept of differentiation with respect to vectors and matrices in order to make the meaning of equations more precise. Clearly, the derivative of a scalar function, such as the Lagrangian, with respect to a column vector is a row vector. For column vectors $\mathbf{a}$ and $\mathbf{b}$, and a function $F(\mathbf{a})$, we have

$$\frac{\partial F}{\partial \mathbf{a}}\mathbf{b} = \left(\frac{\partial F}{\partial \mathbf{a}}\right)^T \cdot \mathbf{b} = \sum \frac{\partial F}{\partial a_i} b_i = (\text{row})(\text{vector}) = (\text{scalar}). \quad (12.2)$$

The equivalent representation of derivatives in terms of matrices is less straightforward. We will typically take derivatives of functions of the type $F(a) = \frac{1}{2}\langle \mathbb{D}a, a \rangle$ for antisymmetric matrices a and a diagonal matrix $\mathbb{D} = \text{diag}(d_1, d_2, d_3)$, having the physical meaning of the inertia matrix. One can readily check that the matrix

$\frac{\partial F}{\partial a} = \mathbb{D}a$ is, in general, not antisymmetric, so it cannot be directly interpreted as a vector. However, for any antisymmetric matrix b, the product $\left\langle \frac{\partial F}{\partial a}, b \right\rangle$ only depends on the antisymmetric part of $\frac{\partial F}{\partial a}$. Thus, the following quantity is readily interpreted as a vector:

$$\frac{\partial F}{\partial \mathbf{a}} = \frac{1}{2}\left[\frac{\partial F}{\partial a} - \left(\frac{\partial F}{\partial a}\right)^T \right]^\vee. \tag{12.3}$$

Because of the awkwardness of the right-hand side of Eq. (12.3), we shall always use vector derivatives (Eq. (12.2)) in the formulas in this chapter.

12.3 Variational Derivation of Rigid Body Equations

As we have seen in Chap. 10, we can derive the equations of motion for a rigid body using the balance of angular momentum in the body frame. In this section, we show how to derive this equation using the variational principle using the Euler-Poincaré variational method.

Let us return to the question of rigid body dynamics and consider a left-invariant Lagrangian $L(\mathbb{R}, \dot{\mathbb{R}})$ with respect to arbitrary rotations of the space. As we mentioned, we can rewrite this Lagrangian as a function of the angular velocity only, i.e., we have $L(\mathbb{R}, \dot{\mathbb{R}}) = \ell\big((\mathbb{R}^T\dot{\mathbb{R}})^\vee\big) = \ell(\mathbf{\Omega})$ for the reduced Lagrangian ℓ defined on *three*-vectors and given by the kinetic energy: $\ell(\mathbf{\Omega}) = \frac{1}{2}\mathbb{I}\mathbf{\Omega} \cdot \mathbf{\Omega}$. How do we write the analogs of the variational principle and the Euler-Lagrange equations for the Lagrangian $\ell(\mathbf{\Omega})$? If we write the variations of the action as

$$\delta \int_{t_0}^{t_1} L(\mathbb{R}, \dot{\mathbb{R}})\mathrm{d}t = \delta \int_{t_0}^{t_1} \ell(\mathbf{\Omega})\mathrm{d}t = \int_{t_0}^{t_1} \frac{\partial \ell}{\partial \mathbf{\Omega}} \cdot \delta\mathbf{\Omega}\mathrm{d}t ,$$

we need to compute the variations $\delta\mathbf{\Omega}$ that are induced by the variations $\delta\Lambda$. Let us define $\Sigma = \mathbb{R}^T\delta\mathbb{R}$, which is also an antisymmetric matrix. This matrix can be associated with a vector $\mathbf{\Sigma} = \Sigma^\vee$. Then, we compute the variations

$$\delta\Omega = \delta\left(\mathbb{R}^T\dot{\mathbb{R}}\right) = \delta\left(\mathbb{R}^T\right)\dot{\mathbb{R}} + \mathbb{R}^T\delta\dot{\mathbb{R}} = -\mathbb{R}^T\delta\mathbb{R}\mathbb{R}^T\dot{\mathbb{R}} + \mathbb{R}^T\delta\dot{\mathbb{R}}^T$$

$$= -\widehat{\Sigma}\widehat{\Omega} + \mathbb{R}^T\delta\dot{\mathbb{R}}$$

$$\frac{\mathrm{d}}{\mathrm{d}t}\widehat{\Sigma} = \frac{\mathrm{d}}{\mathrm{d}t}\left(\mathbb{R}^T\delta\mathbb{R}\right) = \frac{\mathrm{d}}{\mathrm{d}t}\left(\mathbb{R}^T\right)\delta\mathbb{R} + \mathbb{R}^T\delta\dot{\mathbb{R}}$$

$$= -\mathbb{R}^T\dot{\mathbb{R}}\mathbb{R}^T\delta\mathbb{R} + \mathbb{R}^T\delta\dot{\mathbb{R}} = -\widehat{\Omega}\widehat{\Sigma} + \mathbb{R}^T\delta\dot{\mathbb{R}}. \tag{12.4}$$

In Eq. (12.4), we have used the fact that the δ derivative and the time derivative commute. We also used the fact that

$$\frac{d}{dt}\mathbb{A}^{-1} = -\mathbb{A}^{-1}\dot{\mathbb{A}}\mathbb{A}^{-1}, \quad \text{consequently,} \quad \delta\mathbb{A}^{-1} = -\mathbb{A}^{-1}(\delta\mathbb{A})\mathbb{A}^{-1},$$

since the variation δ is, formally, the derivative with respect to some parameter before setting the value of that parameter to 0. Subtracting the equations (Eq. (12.4)) to eliminate the cross-derivatives $\delta\dot{\mathbb{R}}$, we obtain the expression for the variation of $\widehat{\Omega}$ in terms of $\widehat{\Sigma}$ as

$$\delta\widehat{\Omega} = \frac{d}{dt}\widehat{\Sigma} + \left[\widehat{\Omega}, \widehat{\Sigma}\right] \quad \Leftrightarrow \quad \delta\mathbf{\Omega} = \dot{\mathbf{\Sigma}} + \mathbf{\Omega} \times \mathbf{\Sigma}. \tag{12.5}$$

Substitution of Eq. (12.5) into the variational principle, integrating by parts once and using that $\mathbf{\Sigma}(t_0) = \mathbf{\Sigma}(t_1) = 0$ as a consequence of $\delta\Lambda(t_0) = \delta\Lambda(t_1) = 0$, gives

$$\delta\int_{t_0}^{t_1} \ell(\mathbf{\Omega})\, dt = \int_{t_0}^{t_1} \frac{\partial\ell}{\partial\mathbf{\Omega}} \cdot \delta\mathbf{\Omega} dt = \int_{t_0}^{t_1} \frac{\partial\ell}{\partial\mathbf{\Omega}} \cdot \left(\dot{\mathbf{\Sigma}} + \mathbf{\Omega} \times \mathbf{\Sigma}\right)\, dt$$

$$= -\int_{t_0}^{t_1} \left(\frac{d}{dt}\frac{\partial\ell}{\partial\mathbf{\Omega}} + \mathbf{\Omega} \times \frac{\partial\ell}{\partial\mathbf{\Omega}}\right) \cdot \mathbf{\Sigma}\, dt. \tag{12.6}$$

Since $\mathbf{\Sigma}(t)$ is an arbitrary function of time, the equations of motion are

$$\frac{d}{dt}\frac{\partial\ell}{\partial\mathbf{\Omega}} + \mathbf{\Omega} \times \frac{\partial\ell}{\partial\mathbf{\Omega}} = 0 \quad \Rightarrow \quad \frac{d}{dt}\mathbb{I}\mathbf{\Omega} = \mathbb{I}\mathbf{\Omega} \times \mathbf{\Omega}, \tag{12.7}$$

which are exactly Euler's equations for the motion of a rigid body (Eq. (10.19)) we derived before. We would like to draw the attention of the reader to the fact that the function multiplying $\mathbf{\Sigma}$ in Eq. (12.6) is *exactly the angular momentum balance*.

We can also demonstrate how to derive the heavy top equations (Eq. (10.24)) using the Euler-Poincaré variational principle. We need to introduce the variable $\mathbf{\Gamma} = \Lambda^T \mathbf{e}_3$, so

$$\delta\mathbf{\Gamma} = \delta\mathbb{R}^T \mathbf{e}_3 = (\mathbb{R}^T \delta\mathbb{R})^T \mathbb{R}^T \mathbf{e}_3 = -\mathbf{\Sigma} \times \mathbf{\Gamma}. \tag{12.8}$$

Then, we write the variation of the action for Lagrangian $\ell = \ell(\mathbf{\Omega}, \mathbf{\Gamma})$ as

$$\delta\int_{t_0}^{t_1} \ell(\mathbf{\Omega}, \mathbf{\Gamma})dt = \int_{t_0}^{t_1} \frac{\partial\ell}{\partial\mathbf{\Omega}} \cdot \delta\mathbf{\Omega} + \frac{\partial\ell}{\partial\mathbf{\Gamma}} \cdot \delta\mathbf{\Gamma}\, dt$$

$$= \int_{t_0}^{t_1} \frac{\partial\ell}{\partial\mathbf{\Omega}} \cdot \left(\dot{\mathbf{\Sigma}} + \mathbf{\Omega} \times \mathbf{\Sigma}\right) + \frac{\partial\ell}{\partial\mathbf{\Gamma}} \cdot \left(-\mathbf{\Sigma} \times \mathbf{\Gamma}\right)\, dt \tag{12.9}$$

$$= -\int_{t_0}^{t_1} \left(\frac{d}{dt}\frac{\partial\ell}{\partial\mathbf{\Omega}} + \mathbf{\Omega} \times \frac{\partial\ell}{\partial\mathbf{\Omega}} + \mathbf{\Gamma} \times \frac{\partial\ell}{\partial\mathbf{\Gamma}}\right) \cdot \mathbf{\Sigma}\, dt.$$

Taking $\ell = \frac{1}{2}\mathbf{\Omega} \cdot \mathbb{I}\mathbf{\Omega} - mg\mathbf{A} \cdot \mathbf{\Gamma}$, and equating the coefficient multiplying $\mathbf{\Sigma}$ to zero, gives exactly Euler-Poisson equations for the heavy top (Eq. (10.24)). Thus, the advantage of the variational derivation is that the angular and the linear momentum balance are computed automatically through a well-defined procedure, no matter how complex the Lagrangian may be.

12.4 Looking Forward

The variational methods are very powerful, as they allow us to obtain equations of motion and their analysis in a very general setting, using the machinery of variational principles that give equations of motion valid for arbitrary Lagrangians. There is no need to balance forces and torques: variational methods do that for you. Thus, variational methods were applied to a wide variety of problems. Among the problems considered by these methods are satellites with attachments [122], water waves [96, 123], fluid dynamics [96], fluids in flexible tubes (garden hoses) [124, 125], porous media dynamics [126], quantum mechanics [127, 128], liquid crystals [129, 130], molecular dynamics [131], and many others. That list is by no means complete; a reader can easily find many more successes that were accomplished in recent years. In contrast, trying to compute the angular and linear momentum balance equations by equating terms from Newton's laws is often extremely difficult, if not impossible, when the system has nontrivial interactions of several parts. I encourage the reader to use the ideas combining geometry and mechanics whenever possible, and I hope that the ideas from this book were beneficial in showing the reader the advantages of this journey.

Chapter 13
Sample Midterm and Final Exams

13.1 Midterm 1

In Class, Time: 1 h and 20 min

1. **40 points.** A train is moving along a straight line in a prescribed way: in other words, the coordinate $X(t)$ of the train at the time t is given. A train car is carrying a cart with a pendulum. The cart without the pendulum has the mass M and can move along the car without friction. The cart is attached to the train car with a spring of stiffness c and equilibrium length l. There is a pendulum on the cart with a bob of mass m that can swing freely and without friction. All the mass of the pendulum is concentrated in the bob; the rest of the pendulum is massless. The length of the pendulum is l and is constant. The system is illustrated in Fig. 13.1.

   (a) (5 points) Identify the configuration manifold of the system.
   (b) (5 points) Write the coordinates of all the relevant parts of the system contributing to the Lagrangian.
   (c) (5 points) Write the velocities (time derivatives of these coordinates).
   (d) (5 points) Compute the kinetic and potential energies and write the Lagrangian.
   (e) (5 points) Write the Euler-Lagrange equation.
   (f) (5 points) If there are cyclic variables in the Lagrangian, identify them and compute the corresponding conservation law.
   (g) (5 points) Compute the energy as $E = \dot{\mathbf{q}} \cdot \frac{\partial L}{\partial \dot{\mathbf{q}}} - L$. Is the energy E equal to the kinetic plus potential energy?
   (h) (5 points) Compute $\dot{E}$. *Hint* Do not compute that quantity directly from the Euler-Lagrange equations, but rather from the general formula $\dot{E} = -\frac{\partial L}{\partial t}$. Is the energy conserved?

© The Author(s), under exclusive license to Springer Nature Switzerland AG 2025
V. Putkaradze, *A Concise Introduction to Classical Mechanics*,
Surveys and Tutorials in the Applied Mathematical Sciences 16,
https://doi.org/10.1007/978-3-031-84977-0_13

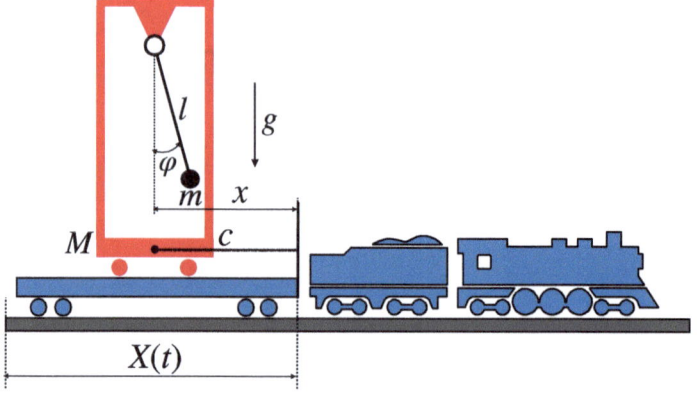

Fig. 13.1 A train carrying a cart with a freely swinging pendulum

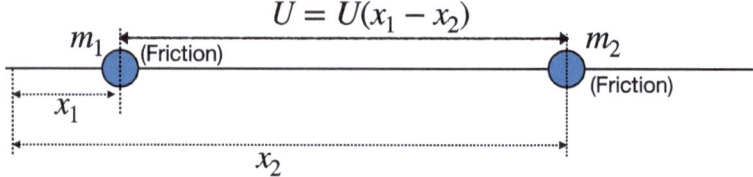

Fig. 13.2 Two beads sliding with friction along a straight wire

2. **25 points**. Two beads are sliding with friction along a straight wire. The potential between the beads depends only on the relative distance between them. The mass of each bead 1, 2 is, correspondingly, m_1 and m_2. The friction force on each bead is $F_i = -\alpha_i \dot{x}_i$, $i = 1, 2$, with $\alpha_i > 0$. The system is illustrated in Fig. 13.2. Note that the direction of friction forces is not shown.

(a) (5 points) Identify the configuration manifold for the system
(b) (5 points) Write the Lagrangian and energy E.
(c) (5 points) Compute the Rayleigh dissipation function.
(d) (5 points) Compute the dissipative forces and write the Euler-Lagrange equations.
(e) (5 points) Compute $\dot{E}$. Is the system dissipative, i.e., is $\dot{E} \leq 0$?

3. **10 points**. Consider the following Lagrangian $L = L(\mathbf{q}, \dot{\mathbf{q}})$, with $\mathbf{q} = (q_1, q_2, q_3)$. (We keep the indices of q's down so as not to confuse them with powers):

$$L = \frac{\dot{q}_1^6 + \dot{q}_2^4 + \dot{q}_3^2}{q_1^6 + q_2^4 + q_3^2} \tag{13.1}$$

(a) (5 points) Show that there is a one-parameter transformation $q_i \rightarrow e^{\gamma_i \epsilon} q_i$ that leaves the Lagrangian defined in Eq. (13.1) invariant.
Hint Set, for example, $\gamma_3 = 1$ and compute γ_1 and γ_2.

(b) (5 points) Compute Noether's integral corresponding to that transformation.

4. **Extra credit: 5 points total**: A tennis ball is thrown through the air and is flying horizontally with respect to the ground. The ball is covered by honey and encounters a cloud of flies. The flies like the honey-covered ball, and some of them land on it, and once they land on the ball, they get stuck. In this problem, all air resistance is neglected.

The flies are hovering while being approximately stationary in the air. When they land on the ball, at the moment of impact, the flies are also stationary with respect to the ground. The mass of the ball $m(t)$ changes due to the mass of the flies.

(a) (3 points) Using the conservation of momentum, find the law connecting between the mass $m(t)$ at time t and the velocity of the ball $v(t)$, given that the initial velocity with respect to the ground is v_0 and the initial mass of the ball is m_0.

(b) (2 points) The amount of flies landing on the ball is proportional to the ball's velocity: $\dot{m} = \alpha v$ where $\alpha > 0$ is some constant. Using the relationship between $m(t)$ and $v(t)$ found in part (a), write the differential equation satisfied by $m(t)$ and solve it to find the mass of the ball at time t.

13.2 Midterm 2

In-Class, 1 h and 20 min

1. **30 points.** A wheel of mass m, radius R, and moment of inertia I is attached at the center to the wall with a spring of equilibrium length d_0 and stiffness c. The wheel is rolling without slipping on the surface. A pendulum of length l and bob's mass m is attached to the center of the wheel O and can swing freely without friction. The problem is illustrated in Fig. 13.3.

(a) (5 points) Identify the configuration manifold of the system.

(b) (5 points) Write the coordinates for all the relevant parts of the system contributing to the Lagrangian.
Hint It is easiest to choose the angle of the wheel's rotation θ or coordinate of the wheel's center x so $\theta = 0$ (resp. $x = 0$) corresponds to the position of the spring in equilibrium.

(c) (5 points) Write the velocities (time derivatives of these coordinates). Compute the kinetic energy.

(d) (5 points) Compute the potential energy and write the Lagrangian.

(e) (5 points) Find the equilibrium points; compute the mass matrix $\mathbb{M}$ and the Hessian of the potential energy $\mathbb{K}$ at the equilibrium point.

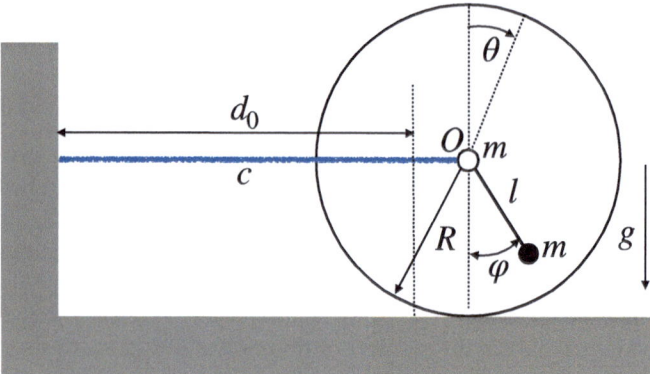

Fig. 13.3 A pendulum is attached to the center of the wheel at point O, and the wheel is attached to the wall at the center (point O) with the spring of stiffness c and equilibrium length d_0

(f) (5 points) Compute the eigenvalues of the linear stability matrix $\Lambda = M^{-1}K$. Don't compute the eigenvectors to save time. State whether the system is linearly stable or unstable.

Hint The inverse of a 2×2 matrix M is computed as

$$M = \begin{pmatrix} a & b \\ c & d \end{pmatrix}, \quad \Rightarrow \quad M^{-1} = \frac{1}{\det M} \begin{pmatrix} d & -b \\ -c & a \end{pmatrix}. \tag{13.2}$$

2. **20 points**. Consider the system from Problem 1, illustrated in Fig. 13.3:

(a) (10 points) Using the Lagrangian computed in Problem 1, determine momenta $\mathbf{p} = \frac{\partial L}{\partial \dot{\mathbf{q}}}$ and express the velocities $\dot{\mathbf{q}}$ as the function of momenta and coordinates. You will find Eq. (13.2) useful here.
(b) (5 points) Compute the Legendre transform and write the Hamiltonian.
(c) (5 points) Write Hamilton's canonical equations for the evolution of $(\mathbf{p}, \mathbf{q})$ using the Hamiltonian that you found.

3. **15 points**. Consider the following vector fields X and Y and one-form θ:

$$X = \xi^1 \frac{\partial}{\partial \xi^1} - \xi^3 \frac{\partial}{\partial \xi^2} + \xi^2 \frac{\partial}{\partial \xi^3},$$

$$Y = \xi^2 \frac{\partial}{\partial \xi^1} - \xi^1 \frac{\partial}{\partial \xi^2} + \xi^3 \frac{\partial}{\partial \xi^3}, \tag{13.3}$$

$$\theta = \xi^2 d\xi^1 + \xi^3 d\xi^2 + \xi^1 d\xi^3.$$

(a) (5 points) Compute $[X, Y]$.
(b) (5 points) Compute $i_X \theta$ and $d(i_X \theta)$.

(c) (5 points) Compute $d\theta$ and $d\theta(X, Y)$. You do not have to simplify the answer to $d\theta(X, Y)$.

4. **15 points.** Consider the generating function of type 2, with $(\mathbf{q}, \mathbf{P}, t)$ being independent variables:

$$F_2(\mathbf{q}, \mathbf{P}, t) = \sum_{i=1}^{n} a_i e^{b_i P_i} \sin q^i t,$$

where both $\mathbf{q} = (q^1, \dots, q^n)$ and $\mathbf{P} = (P_1, \dots, P_n)$ are n-dimensional, and $a_i > 0$ and $b_i > 0$ are some given constants.

(a) (5 points) Compute the variables $(\mathbf{p}, \mathbf{Q})$ as functions $(\mathbf{q}, \mathbf{P})$ according to that generating function.
(b) (5 points) Find explicitly the transformation $\mathbf{Q}(\mathbf{q}, \mathbf{p}, t)$, $\mathbf{P}(\mathbf{q}, \mathbf{p}, t)$.
(c) (5 points) Given the Hamiltonian

$$H(\mathbf{p}, \mathbf{q}, t) = -\frac{1}{t} \sum_{i=1}^{n} p_i q_i,$$

find the new Hamiltonian $K(\mathbf{P}, \mathbf{Q}, t)$ in the new variables.

Note Remember that the variables $(\mathbf{p}, \mathbf{Q})$, and the new Hamiltonian K are computed according to

$$p_\alpha = \frac{\partial F_2}{\partial q^\alpha}, \quad Q_\alpha = \frac{\partial F_2}{\partial P_\alpha}, \quad K = H + \frac{\partial F_2}{\partial t}.$$

13.3 Final Exam

Take-Home, 24 h

1. **30 points.** A helical wire in the initial state is described by the coordinates:

$$\mathbf{r}_0 = (x_0, y_0, z) = (a \cos(\gamma z), a \sin(\gamma z), z),$$

where $a > 0$ and $\gamma > 0$ are some constants. That helical wire can rotate about its axis (z) without friction, with the rotation causing elastic torque in the joint supporting the axis. The helical wire's moment of inertia of rotation about the axis is I. The mass of the helical wire is not important since the helix only rotates about its axis and doesn't move in space, and the center of mass of the wire is supposed to be on the axis. If the helical wire is rotated by the angle φ, that rotation results in the potential $U_e(\varphi)$. A small bead of mass m can move

Fig. 13.4 Setup of a helix on a plane. The helix (blue line) can rotate about its axis (dashed black line), and the bead (red dot) can move along the helix

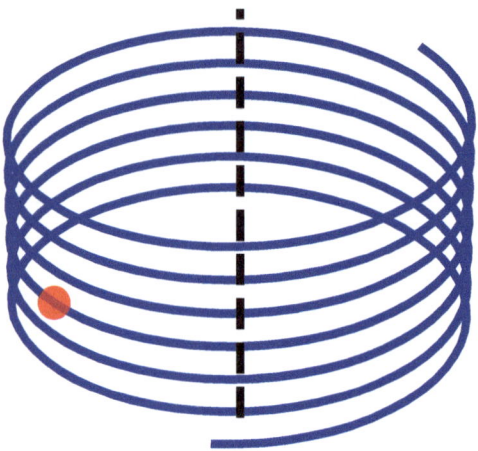

freely (without friction) along the helix. External potential $U_p(z)$ is acting on the bead itself. The system is illustrated in Fig. 13.4.

(a) **(5 pts)** Find the configuration manifold of the system.
(b) **(5 pts)** Find the coordinates of the bead, including the rotation of the helix about the axis by the angle φ.
(c) **(5 pts)** Find the kinetic energy of the system.
(d) **(5 pts)** Compute the Lagrangian of the system and compute the Euler-Lagrange equations for a general $U_e(\varphi)$ and $U_p(z)$.
(e) **(5 pts)** Take the potential energy to be

$$U_p(z) = \frac{1}{2}\alpha z^2, \quad U_e(\varphi) = \frac{1}{2}\beta\varphi^2 .$$

For Euler-Lagrange equations with these potentials, you should get a linear system of equations. Find the corresponding eigenvalues (as in linear stability questions) and conclude whether the system is stable or unstable. You don't have to find the eigenvectors.
(f) **(5 pts)** Find the energy $E = \dot{\mathbf{q}} \cdot \frac{\partial L}{\partial \dot{\mathbf{q}}} - L$ for arbitrary potential energies $U_e(\varphi)$ and $U_p(z)$. Is the energy E conserved?

2. **20 points.** The same as in the previous problem, relating to the motion of a helix, only the rotation angle of the helix $\varphi(t)$ about its axis is now prescribed and is accomplished by an external mechanism (e.g., a controlled motor). There is still an external potential $U_p(z)$ acting on the bead itself as before, but, of course, there are no potential $U_e(\varphi)$ associated with rotation—that potential is compensated for by the control mechanism rotating the helical wire.

(a) **(5 pts)** Find the configuration manifold of the system and the coordinates of the bead, including the rotation of the helix about the axis by the prescribed angle $\varphi(t)$.

(b) **(5 pts)** Find the kinetic energy of the system, compute the Lagrangian of the system, and compute the Euler-Lagrange equations for a general $U_p(z)$ and a general prescribed $\varphi(t)$.

(c) **(5 pts)** Take $\varphi(t) = A \cos \Omega t$ and the potential energy to be

$$U_p(z) = \frac{1}{2}\beta z^2.$$

Find the general solution of the Euler-Lagrange equations. Are there any values of parameters (A, Ω) causing the solutions to be unbounded in time?

(d) **(5 pts)** For an arbitrary potential $U_p(z)$, compute the energy $E = \dot{\mathbf{q}} \cdot \frac{\partial L}{\partial \dot{\mathbf{q}}} - L$. Is that energy conserved?

3. **40 points** *Degenerate Lagrangians.* Consider $\mathbf{q} \in \mathbb{R}^n$ and an (unphysical) degenerate Lagrangian:

$$L = \frac{1}{2}\mathbf{q} \cdot \mathbb{M}\dot{\mathbf{q}} - U(\mathbf{q}) = \frac{1}{2}q^i M_{ij}\dot{q}^j - U(\mathbf{q}), \quad M_{ij} = -M_{ji}, \quad (13.4)$$

so $\mathbb{M}^T = -\mathbb{M}$. Consider also the regularization of this Lagrangian using a small parameter $\epsilon \ll 1, \epsilon > 0$:

$$L_\epsilon = \frac{\epsilon}{2}|\dot{\mathbf{q}}|^2 + L = \frac{\epsilon}{2}|\dot{\mathbf{q}}|^2 + \frac{1}{2}\mathbf{q} \cdot \mathbb{M}\dot{\mathbf{q}} - U(\mathbf{q}) = \frac{\epsilon}{2}|\dot{\mathbf{q}}|^2 + \frac{1}{2}q^i M_{ij}\dot{q}^j - U(\mathbf{q}), \quad (13.5)$$

where again $\mathbb{M}^T = -\mathbb{M}$ or, equivalently, $M_{ij} = -M_{ji}$ like in Eq. (13.4). The Lagrangian L given by Eq. (13.4) has a degenerate mass matrix, and no Legendre transform of that Lagrangian exists; on the contrary, L_ϵ given by Eq. (13.5) is regular, and Hamiltonian, linear stability, etc. can be computed. In this problem, we are going to investigate the difference between the dynamics corresponding to the singular L and regularized L_ϵ Lagrangians.

(a) **(5 pts)** Show that any odd-dimensional antisymmetric matrix $\mathbb{M}$ is singular and therefore non-invertible, i.e., $\det \mathbb{M} = 0$.
Note In what follows, we are interested in non-singular matrices $\mathbb{M}$, so we will assume that the dimension of the configuration manifold is even, i.e., $n = 2k$. One could, in principle, work with singular $\mathbb{M}$ as well, but that problem is much more complex than the one considered here, as the Euler-Lagrange equations become algebraic-differential rather than purely differential equations.

(b) **(5 pts)** Find the (formal) Euler-Lagrange equations corresponding to the Lagrangian (Eq. (13.4)).

(c) **(5 pts)** Show that the Euler-Lagrange equations, in this case, only contain first-order time derivatives in $\mathbf{q}$ and not second-order as usual. Outline what initial conditions you would need to solve the Euler-Lagrange equations given by L in Eq. (13.4). Show also that $U(\mathbf{q})$ is preserved on solutions, and thus all solutions of these Euler-Lagrange equations belong to the manifold $U(\mathbf{q}) =$const.

(d) **(5 pts)** Consider the configuration manifold to be the two-dimensional space $\mathbb{R}^2$ with $\mathbf{q} = (q^1, q^2)$. Consider the matrix $\mathbb{M}$ given by

$$\mathbb{M} = \begin{pmatrix} 0 & 1 \\ -1 & 0 \end{pmatrix} \tag{13.6}$$

and the potential energy to be

$$U(\mathbf{q}) = f(\rho), \quad \rho := \frac{(q^1)^2}{a^2} + \frac{(q^2)^2}{b^2}, \tag{13.7}$$

where $f(\rho)$ is some smooth function of its argument. Find the solution $\mathbf{q}(t) = (q^1(t), q^2(t))$ explicitly as a function of time t. The final answer will, of course, depend on the function $f(\rho)$.

(e) **(5 pts)** Now, take L_ϵ defined by Eq. (13.5). Find the Euler-Lagrange equations for L_ϵ for arbitrary $U(\mathbf{q})$. Note that these equations are second order and not first order as defined by L.

(f) **(5 pts)** Find the energy function $E = \dot{\mathbf{q}} \cdot \frac{\partial L}{\partial \dot{\mathbf{q}}} - L$ for both Eqs. (13.4) and (13.5). Is that energy conserved in both cases?

(g) **(5 pts)** Now, take a particular case of the potential energy (Eq. (13.7)) with $f(\rho) = K\rho$ (you can assume $K > 0$). Also, take $a = b = 1$ and, as before, $\mathbb{M}$ given by Eq. (13.6). Write the Euler-Lagrange equation defined by Eqs. (13.4) and (13.5) for that potential energy. Write the solutions for both of these Euler-Lagrange equations. Explain what happens to the solutions of the Euler-Lagrange equations for L_ϵ when $\epsilon \to 0$.
Hint In the regularized case (Eq. (13.5)), look for solutions $\mathbf{q} = \mathbf{q}_0 e^{\lambda t}$ (or $\mathbf{q} = \mathbf{q}_0 e^{i\omega t}$ if you prefer) and find λ (resp. ω) from the Euler-Lagrange equations from the conditions of vanishing determinants for some matrix.

(h) **(5 pts)** Sketch the solutions of the Euler-Lagrange equations for both cases as obtained in the previous part, and explain how solutions for the Euler-Lagrangian equations given by the regularized Lagrangian L_ϵ from Eq. (13.5) relate to the solutions of the Euler-Lagrange equations given by the singular Lagrangian L defined in (13.4).

4. **25 points** *Rayleigh's dissipation function.* Consider a particle in a three-dimensional space $\mathbf{q} = (x, y, z) \in \mathbb{R}^3$. The particle can move while being constrained to the surface $z = f(x, y)$ while exposed to the potential $U(x, y, z)$. The friction force is proportional to the particle's velocity in the full space $\mathbb{R}^3$, i.e., $\mathbf{F}_f = -\alpha \dot{\mathbf{q}}$ for some $\alpha > 0$.

(a) **(5 pts)** Find the configuration manifold for the system.

(b) **(5 pts)** Find the coordinates, their time derivatives, the kinetic energy, and the Lagrangian L.

(c) **(5 pts)** Compute the friction force and Raleigh's dissipation function.

(d) **(5 pts)** Write the Euler-Lagrange equations including the friction force. Do not simplify the equations.

(e) **(5 pts)** Show that for a rotationally symmetric surface $f = f(r)$, $r = \sqrt{x^2 + y^2}$, gravitational potential $U = kz$, and the absence of friction $\alpha = 0$, there is Noether's integral arising from the symmetry of rotation about the vertical axis, and that integral has the physical meaning of z-component of the angular momentum.

5. **10 points** Consider a canonical Hamiltonian system where $\mathbf{p}$ and $\mathbf{q}$ are n-dimensional vectors.

(a) **(5 pts)** Compute the Poisson bracket $\{\varphi, \psi\}$ of the functions $\varphi = f(\xi)$, and $\psi = g(\xi)h(\eta)$, where $\xi = |\mathbf{q} + \mathbf{p}|^2$ and $\eta = \mathbf{a} \cdot \mathbf{q} + \mathbf{b} \cdot \mathbf{p}$. Here, f, g, and h are some smooth functions of their arguments, and $\mathbf{a}$ and $\mathbf{b}$ are some fixed n-dimensional vectors.

(b) **(5 pts)** Find all the values of $\mathbf{a}$ and $\mathbf{b}$ such that $\{\varphi, \psi\} = 0$, if they exist. If these values of $\mathbf{a}$ and $\mathbf{b}$ don't exist, explain why.

6. **15 points** Two beads of mass m can slide without friction on a hoop of radius R. The beads are connected with a spring of equilibrium length $l < 2R$ and stiffness c. The hoop is rotating with a constant angular velocity ω. The two beads are also acted upon by gravity; assume that the spring is massless. The system is shown in Fig. 13.5.

(a) **(5 pts)** Find the configuration manifold of the system.

(b) **(5 pts)** Find the coordinates, velocities, and kinetic and potential energies.

(c) **(5 pts)** Perform the Legendre transform and compute the Hamiltonian of the system. Write out the canonical Hamilton's equations of motion.

Fig. 13.5 Two beads on a rotating hoop connected with a spring

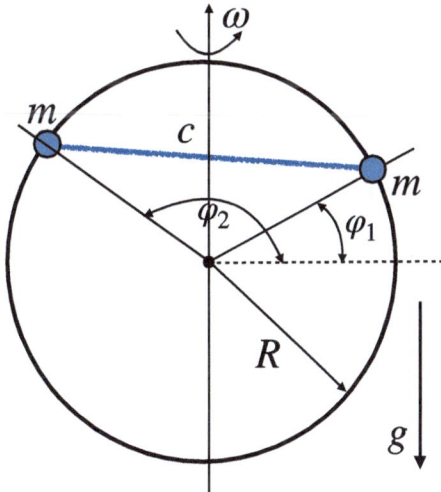

7. **20 points**. In each of the problems below, the type of canonical transformations and the generating functions are given. Find the expressions of the remaining variables, and then find the transformation $\mathbf{Q} = \mathbf{Q}(\mathbf{p}, \mathbf{q}, t)$ and $\mathbf{P} = \mathbf{P}(\mathbf{p}, \mathbf{q}, t)$. Find the transformed Hamiltonian $K(\mathbf{P}, \mathbf{Q}, t)$ in the new variables. Everywhere below, all vectors $\mathbf{q}, \mathbf{p}$, etc. are n-dimensional.

(a) **(10 pts)** Type 1 transformation,

$$F_1(\mathbf{q}, \mathbf{Q}, t) = \sum_{i=1}^{n} a_i \cos(Q^i t + q^i), \quad H = \sum_{i=1}^{n} e^{Q^i t + q^i}.$$

(b) **(10 pts)** Type 4 transformation: consider an $n \times n$ matrix $\mathbb{A}$ that is symmetric $\mathbb{A}^T = \mathbb{A}$ and invertible, so $\mathbb{B} = \mathbb{A}^{-1}$ exists, with the (i, j)-th entry of $\mathbb{A}$ being A_{ij} and (i, j)-th entry of $\mathbb{B}$ being B_{ij}. Then, take the generating function and the Hamiltonian to be

$$F_4(\mathbf{p}, \mathbf{P}, t) = \sum_{(i,j)=1}^{n} A_{ij} \log(p_i t) \log(P_j t) = \log(\mathbf{p}t) \cdot \mathbb{A} \log(\mathbf{P}t),$$

$$H = \sum_{i=1}^{n} q^i p_i \log(p_i t).$$

8. **45 points** *Hamilton-Jacobi equation*. For each of the following Hamiltonians, write the corresponding Hamilton-Jacobi equation. Then, find a solution (integral) of that Hamilton-Jacobi equation. Using that solution, find the general solution of the canonical Hamilton's equations in quadratures or as implicit functions. All indices of (p_i, q_i) are down to avoid possible confusion with the powers.

(a) **15 points**

$$H = \frac{\cos(p_1 + q_1) + \cos(p_2 + q_2)}{\sin(p_1 + q_1) + \sin(p_2 + q_2)} \cos t.$$

(b) **15 points**

$$H = \cos[p_1 + q_1 \sin(p_2 + q_2)] \sin t.$$

(c) **15 points**

$$H = p_1 + q_1^2 \frac{p_2 + p_3}{q_2 + q_3}.$$

9. **10 points** For the heavy top equations,

$$\mathbb{I}\dot{\boldsymbol{\Omega}} = -\boldsymbol{\Omega} \times \mathbb{I}\boldsymbol{\Omega} + mg\boldsymbol{\Gamma} \times \mathbf{A}, \quad \dot{\boldsymbol{\Gamma}} = -\boldsymbol{\Omega} \times \boldsymbol{\Gamma}. \tag{13.8}$$

All notation as in the notes: $\boldsymbol{\Omega}$ is the angular velocity in the body frame, $\boldsymbol{\Gamma}$ is the vertical vector as seen from the body, $\mathbb{I}$ is the moment of inertia, and $\mathbf{A}$ is the fixed vector (in the body frame) from the fixed point to the center of mass. Compute the evolution of the following quantities $F(\boldsymbol{\Omega}, \boldsymbol{\Gamma})$, i.e., $\frac{dF}{dt}$ when $(\boldsymbol{\Omega}, \boldsymbol{\Gamma})$ satisfy Eq. (13.8):

(a) **5 points.** $F = |\mathbf{A} \times \mathbb{I}\boldsymbol{\Omega}|^2$.
(b) **5 points.** $F = \boldsymbol{\Gamma} \cdot (\mathbf{A} \times \mathbb{I}\boldsymbol{\Omega})$.

10. **20 points** A skater is moving on a plane in a potential field $U(x, y)$. The angle of the skate with respect to the x-axis is θ. The mass of the skate is m, and the moment of inertia of rotation about the axis normal to the plane (i.e., angle θ), with respect to the center of mass, is J. The nonholonomic skating constraint is given by

$$\dot{x}\sin\theta - \dot{y}\cos\theta = 0$$

The system is illustrated in Fig. 13.6.

(a) **(10 pts)** Using Lagrange-d'Alembert's principle, find the equations of motion for the skate.
(b) **(5 pts)** Using the equations of motion you have derived, prove that the total energy is conserved. The energy, in this case, is just kinetic plus potential energy.
(c) **(5 pts)** Find the equation of motion for the component of momentum parallel to the skate, i.e., compute $m\dot{v}$ where $v = \dot{x}\cos\theta + \dot{y}\sin\theta$. Note that also $\dot{x} = v\cos\theta$ and $\dot{y} = v\sin\theta$, and write the system of equations for (v, x, y, θ) for an arbitrary potential energy $U(x, y)$.

Fig. 13.6 Setup of a model skater on a plane

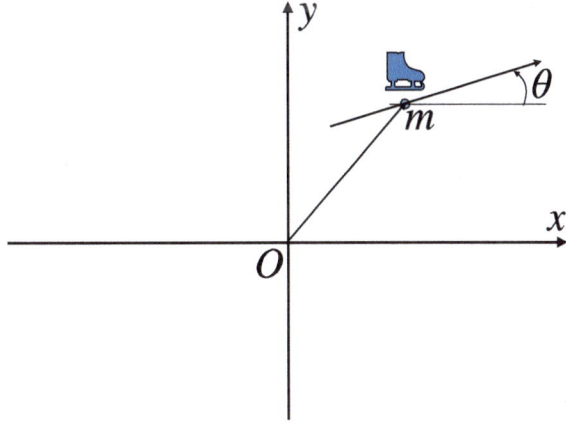

Appendix A
Solutions to Selected Practice Problems

A.1 Chapter 1

Problem 3 Suppose the mass of a rocket at time t is $M(t)$ and the velocity in an inertia frame is $V(t)$. Between the time t and $t + \Delta t$, the mass ΔM leaves the rocket through the nozzle, having the velocity U with respect to the nozzle. The momentum of the jet fuel is, therefore, $\Delta M(V - U)$. Then, the momentum of the rocket at time t is $M(t)V(t)$. The rocket mass, which is now $M - \Delta M$, yields the rocket's momentum at time $t + \Delta t$ to be $(M - \Delta M)(V + \Delta V)$. The conservation of momentum gives

$$(M - \Delta M)(V + \Delta V) + \Delta M(V - U) = MV,$$
$$\Rightarrow M\Delta V - U\Delta M - \Delta M\Delta V = 0. \tag{A.1}$$

Remembering that $\Delta M = -\dot{M}\Delta t$ (the mass is decreasing if the amount of ΔM is taken out of the rocket), $\Delta V = \dot{V}\Delta t$, expanding Eq. (A.1) and dropping Δt^2 term, we get

$$\dot{V}M = -\dot{M}U \quad \Rightarrow \quad \dot{V} = -U\frac{\dot{M}}{M} \quad \Rightarrow \quad V = V_0 - U \log \frac{M(t)}{M(0)}. \tag{A.2}$$

So if the rocket ends up with the final mass $M_f = m$, then the final velocity is

$$V_f = V_0 + U \log \frac{M_0}{m}. \tag{A.3}$$

Problem 4 Let us consider the center of mass of the bad guy of mass M, instantly acquiring horizontal velocity V_* by absorbing the bullet of mass m. Since the bad guy is instantly incapacitated, there is no longer support from his legs, so the center

V. Putkaradze, *A Concise Introduction to Classical Mechanics*,
Surveys and Tutorials in the Applied Mathematical Sciences 16,
https://doi.org/10.1007/978-3-031-84977-0

of mass, now having a mass $M + m$, is flying horizontally with the velocity V_* from the initial height $h = 1$m. The center of mass is traveling the distance $L = 1$m. The time to fall from height h is given by $\Delta t = \sqrt{2h/g}$ g being the acceleration of gravity), and so the velocity V_* can be estimated as $V_* = L/\Delta t \simeq 3$m/s.

One can say conjecture that the system good guy+bad guy+bullet is closed, so the recoil from the bullet, in the first approximation, should throw the good guy approximately the same distance as the bad guy. However, this argument is not really precise as there is the interaction of both the good guy and the bad guy with the ground before and after the shot. A more precise calculation considers the system bad guy+bullet before and after impact, which is closed.

If the velocity of the bullet is V_0, the conservation of momentum leads to

$$(M + m)V_* = mV_0, \quad \Rightarrow \quad m = M\frac{V_*}{V_0 - V_*} \text{ or } V_0 = \frac{M + m}{m}V_*. \tag{A.4}$$

Taking a bullet from Colt .45 in the late 1800 s (about 1000 ft/s, or 300 m/s), y'all get $m = 0.8$ kg from Eq. (A.4) (which is more like a cannonball than a bullet). If y'all take the regular mass of the bullet of around 16 g for a period pistol, then y'all get $V_0 \simeq 5000 \times V_* \simeq 15$ km/s (the bullet could go to space).

A.2   Chapter 2

Problem 1 Suppose $V_0 = Kr^p$, $p > 0$. The effective potential and the effective energy are

$$W_e(r) = \frac{l^2}{2mr^2} + Kr^p, \quad E_l = \frac{1}{2}m\dot{r}^2 + \frac{l^2}{2mr^2} + Kr^p. \tag{A.5}$$

The solution curves on the $(r, \dot{r})$ plane are given by the level set $E = $const. All the cases $K > 0$ and $p > 0$ are qualitatively the same. The important feature of the energy W_e defined by Eq. (A.5) is that $E \to +\infty$ when $r \to 0$ or $r \to \infty$. Thus, all the trajectories, for all values of energy $E_l = E$, are closed. Examples of the trajectories are illustrated in Fig. A.1.

Problem 2 Suppose $V_0 = -Kr^{-p}$, $p > 0$. The effective potential and the effective energy are

$$W_e(r) = \frac{l^2}{2mr^2} - Kr^{-p}, \quad E_l = \frac{1}{2}m\dot{r}^2 + \frac{l^2}{2mr^2} - Kr^{-p}. \tag{A.6}$$

One can see that $W_e(r)$ has a minimum for $W_e = W_*$ for $0 < p < 2$, and $W_e(r) \to \infty$ as $r \to 0$, $W_e(r) \to 0$ as $r \to \infty$. Thus, the solutions with $E_l > 0$ are unbounded, and solutions with $W_* < E_l < 0$ are bounded. Examples of the trajectories are illustrated in Fig. A.2. Finally, when $p = 2$, the powers of r match

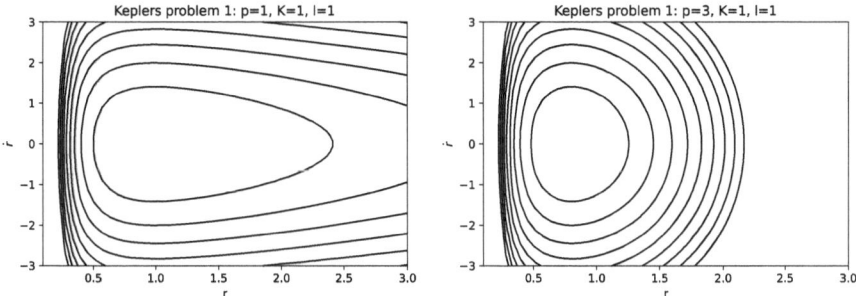

Fig. A.1 Level sets in the $(r, \dot{r})$ plane for $K = 1, l = 1$, and $p = 1$ (left) and $p = 3$ (right)

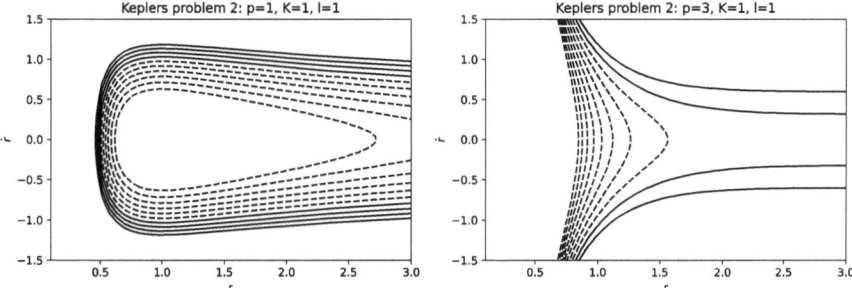

Fig. A.2 Level sets in the $(r, \dot{r})$ plane for $K = 1, l = 1$, and $p = 1$ (left) and $p = 3$ (right)

exactly in the effective potential energy. If $K > l^2/(2m)$, the potential energy is $W_e = -K_* r^{-2}$, so the trajectories are closed for $E < 0$ and unbounded for $E > 0$. If $K < l^2/(2m)$, then $W_e = K_* r^{-2}$, so all trajectories are unbounded. Finally, if $K = l^2/(2m)$, then the potential energy vanishes exactly, so the trajectories are simply straight lines $\dot{q} = $ const.

Problem 4 A direct substitution of this Lagrangian into Euler-Lagrange equations gives an impossible equation $1 = 0$. This outcome is an example of *degenerate Lagrangians*, occurring when the second derivative with respect to $\dot{q}$ vanishes. In general, when the Lagrangian $L(\mathbf{q}, \dot{\mathbf{q}}, t)$ is such that the Hessian matrix $\frac{\partial^2 L}{\partial \dot{\mathbf{q}} \partial \dot{\mathbf{q}}}$ is degenerate (has an eigenvalue 0). In that case, Euler-Lagrange equations will be hard or impossible to solve. These particular cases of degenerate Lagrangians happen physically when kinetic energy terms from some of the masses are neglected because the masses are assumed to be small.

In some cases of degenerate Lagrangians, we do not get impossible equations, and one can achieve some progress. For example, there are some Lagrangians that are first order in $\dot{\mathbf{q}}$ that, upon the substitution into the Euler-Lagrange equations, give *first-order* equations. In that case, the initial velocity $\dot{q}(0)$ cannot be treated as a specified initial condition but needs to be chosen in such a way that it satisfies the Euler-Lagrange equation. An example of such a Lagrangian is given in the sample final exam, Problem 3. That problem also shows the limiting process of taking the mass to 0, which makes the Lagrangian degenerate.

A.3   Chapter 3

Problem 1

1. Configuration manifold could be taken as $\mathbb{R}$ (position of the ball), or $Q = S^1$ (angle of the rotation of the ball). We take the last choice, $Q = S^1$, with the rotation of the ball denoted by φ. The hamster is moving along the ball with the angular velocity $\Omega = V/R$ (or $\Omega = V/(R - h)$).
2. The coordinates of the ball and the hamster are (if we take φ to be measured from the negative axis, so $\varphi = 0$ is down):

   (a) $\mathbf{x}_B = (R\varphi, 0)$,
   (b) $\mathbf{x}_H = (R\varphi + (R - h)\sin(\Omega t + \varphi), -(R - h)\cos(\Omega t + \varphi))$.

3. The velocities of the ball and the hamster are as follows:

   (a) $\dot{\mathbf{x}}_B = (R\dot{\varphi}, 0)$.
   (b) $\dot{\mathbf{x}}_H = (R\dot{\varphi} + (R - h)\cos(\Omega t + \varphi))(\Omega + \dot{\varphi}), (R - h)\sin(\Omega t + \varphi)(\Omega + \dot{\varphi}))$.

4. Calculation of the Lagrangian:

   (a) Kinetic energy of the ball:

   $$T_B = \frac{1}{2}\left(M\dot{x}^2 + I\dot{\varphi}^2\right) = \frac{1}{2}\left(MR^2 + I\right)\dot{\varphi}^2.$$

   (b) Kinetic energy of the hamster $T_H = \frac{1}{2}m\,|\dot{\mathbf{x}}_H|^2$ is computed as

   $$T_H = \frac{1}{2}m\,|\dot{\mathbf{x}}_H|^2$$
   $$= \frac{1}{2}m\left[R^2\dot{\varphi}^2 + (R - h)^2(\dot{\varphi} + \Omega)^2 + 2(R - h)R\dot{\varphi}(\dot{\varphi} + \Omega)\cos(\Omega t + \varphi)\right].$$

   (c) Potential energy only relates to the hamster:

   $$V_h = -mg(R - h)\cos(\Omega t + \varphi).$$

(d) $L = T_B + T_H - V_H$:

$$
L = \frac{1}{2}\left(MR^2 + I\right)\dot{\varphi}^2
$$

$$
+ \frac{1}{2}m\left[R^2\dot{\varphi}^2 + (R-h)^2(\dot{\varphi}+\Omega)^2 + 2(R-h)R\dot{\varphi}(\dot{\varphi}+\Omega)\cos(\Omega t + \varphi)\right]
$$

$$
+ mg(R-h)\cos(\Omega t + \varphi).
$$

$$(A.7)$$

5. Clearly, the Lagrangian (Eq. (A.7)) written above depends on time and so is very uncomfortable to work with. Notice, however, if we take $q = \varphi + \Omega t$ (which happens to be the angle of the hamster with respect to the vertical), we simplify Eq. (A.7) as

$$
q = \varphi + \Omega t, \quad \dot{q} = \dot{\varphi} + \Omega
$$

$$
L = \frac{1}{2}\left(MR^2 + I\right)(\dot{q} - \Omega)^2
$$

$$
+ \frac{1}{2}m\left[R^2(\dot{q}-\Omega)^2 + (R-h)^2\dot{q}^2 + 2(R-h)R(\dot{q}-\Omega)\dot{q}\cos q\right]
$$

$$
- mg(R-h)\cos q.
$$

$$(A.8)$$

6. Now, the Lagrangian (Eq. (A.8)) doesn't depend on time. We can write Euler-Lagrange equations for that Lagrangian as

$$
\frac{d}{dt}\left[(MR^2+I)(\dot{q}-\Omega) + m\left(R^2(\dot{q}-\Omega)\right.\right.
$$

$$
\left.\left.+(R-h)^2\dot{q} + R(R-h)(2\dot{q}-2\Omega)\cos q\right)\right]
$$

$$(A.9)$$

$$
+ m\sin q\left(g(R-h) + 2(R-h)R(\dot{q}-\Omega)\dot{q}\right) = 0.
$$

7. Since the Lagrangian (Eq. (A.8)) doesn't depend on time, the energy E, defined as

$$
E = \dot{q}\frac{\partial L}{\partial \dot{q}} - L = \dot{q}\left[(MR^2+I)(\dot{q}-\Omega) + m\left(R^2(\dot{q}-\Omega)\right.\right.
$$

$$
\left.\left.+(R-h)^2\dot{q} + R(R-h)(2\dot{q}-2\Omega)\cos q\right)\right] - L,
$$

$$(A.10)$$

is conserved. This energy E is not equal to kinetic plus potential energy; it is something else. Upon some algebra, the energy is written as

$$E = \frac{1}{2}\dot{q}^2 [A + B \cos q] - mg(R - h) \cos q + \text{const}, \qquad (A.11)$$

where A, B are some positive constants depending on the parameters.

8. Steady states of Euler-Lagrange equations (Eq. (A.9)) are obtained by setting all time derivatives in Eq. (A.9) to zero, leading to $\sin q = 0$, or $q = 0$, or $q = \pi$.

9. To find the stability of these points, one could, of course, find the linearization (as is explained in the following chapters), but it is awkward and unnecessary. Instead, we use the energy method, where we notice that the solution in $(q, \dot{q})$ plane is given by $E = const$.

   We note that for $q = 0$ (hamster is at the bottom), energy is locally given as $E = \tilde{A}\dot{q}^2/2 + \tilde{C}q^2/2$. The solution curves, locally, are ellipses, so the solution is stable. On the other hand, for $q = \pi$—the hamster is on the top (how would it stay there is another question), the energy is locally given as $E = \tilde{A}\dot{q}^2/2 - \tilde{C}q^2/2$. The solutions, locally, are hyperbolas (or straight lines—separatrices). In any case, the solution with the hamster on top is unstable (as you may have guessed).

Problem 2 The solution proceeds the same way, except that there are two additional weights: the cat and the hamster.

1. Configuration manifold could be again as $\mathbb{R}$ (position of the ball), or $Q = S^1$ (angle of the rotation of the ball). We use the last option, $Q = S^1$, with the rotation of the ball denoted by φ. The hamster is moving along the ball with the angular velocity $\Omega_H = V/R$ (or $\Omega_H = V/(R - h)$). The cat is moving with the angular velocity $\Omega_C = U/R$ (or $\Omega_C = U/(R - h)$).

2. The coordinates of the ball, the hamster, and the cat are (if we take φ to be measured from the negative axis, so $\varphi = 0$ is down):

   (a) $\mathbf{x}_B = (R\varphi, 0)$,
   (b) $\mathbf{x}_H = (R\varphi + (R - h) \sin(\Omega_H t + \varphi), -(R - h) \cos(\Omega_H t + \varphi))$.
   (c) $\mathbf{x}_C = (R\varphi + (R - h) \sin(\Omega_C t + \varphi), -(R - h) \cos(\Omega_C t + \varphi))$.

3. The velocities of the ball, the cat, and the hamster are (we assume, for simplicity, that the centers of mass of the cat and the hamster are the same distance from the ball/center—these formulas can be easily generalized if needed):

   (a) $\dot{\mathbf{x}}_B = (R\dot{\varphi}, 0)$,
   (b) $\dot{\mathbf{x}}_H = (R\dot{\varphi} + (R-h) \cos(\Omega t + \varphi))(\Omega_H + \dot{\varphi}), (R-h) \sin(\Omega_H t + \varphi)(\Omega_H + \dot{\varphi}))$.
   (c) $\dot{\mathbf{x}}_C = (R\dot{\varphi} + (R-h) \cos(\Omega_C t + \varphi))(\Omega_C + \dot{\varphi}), (R-h) \sin(\Omega_C t + \varphi)(\Omega_C + \dot{\varphi}))$.

4. Calculation of the Lagrangian:

(a) Kinetic energy of the ball:

$$T_B = \frac{1}{2}\left(M\dot{x}^2 + I\dot{\varphi}^2\right) = \frac{1}{2}\left(MR^2 + I\right)\dot{\varphi}^2$$

(b) Kinetic energy of the hamster $T_H = \frac{1}{2}m\,|\dot{x}_H|^2$ is computed as

$$T_H = \frac{1}{2}m_H\Big[R^2\dot{\varphi}^2 + (R-h)^2(\dot{\varphi}+\Omega_H)^2$$
$$+2(R-h)R\dot{\varphi}(\dot{\varphi}+\Omega_H)\cos(\Omega_H t + \varphi)\Big].$$

(c) Kinetic energy of the cat $T_C = \frac{1}{2}m_C\,|\dot{x}_C|^2$ is computed as

$$T_C = \frac{1}{2}m_C\Big[R^2\dot{\varphi}^2 + (R-h)^2(\dot{\varphi}+\Omega_C)^2$$
$$+2(R-h)R\dot{\varphi}(\dot{\varphi}+\Omega_C)\cos(\Omega_C t + \varphi)\Big].$$

(d) Potential energy now relates to the hamster and the cat:

$$V = -m_H g(R-h)\cos(\Omega_H t + \varphi) - m_C g(R-h)\cos(\Omega_C t + \varphi)$$

(e) $L = T_B + T_H + T_C - V$:

$$L = \frac{1}{2}\left(MR^2 + I\right)\dot{\varphi}^2$$
$$+ \frac{1}{2}m_H\Big[R^2\dot{\varphi}^2 + (R-h)^2(\dot{\varphi}+\Omega_H)^2$$
$$+2(R-h)R\dot{\varphi}(\dot{\varphi}+\Omega_H)\cos(\Omega_H t + \varphi)\Big]$$
$$+ \frac{1}{2}m_C\Big[R^2\dot{\varphi}^2 + (R-h)^2(\dot{\varphi}+\Omega_C)^2$$
$$+2(R-h)R\dot{\varphi}(\dot{\varphi}+\Omega_C)\cos(\Omega_C t + \varphi)\Big]$$
$$+ m_H g(R-h)\cos(\Omega_H t + \varphi) + m_C g(R-h)\cos(\Omega_C t + \varphi)$$
$$\text{(A.12)}$$

5. The Lagrangian (Eq. (A.12)) written above depends on time. There is no transformation of variables making it independent of time unless $\Omega_H = \Omega_C$.

6. The Euler-Lagrange equations are written in a standard way:

$$\frac{d}{dt}\frac{\partial L}{\partial \dot{\varphi}} - \frac{\partial L}{\partial \varphi} = 0, \quad \text{where}$$

$$\frac{\partial L}{\partial \dot{\varphi}} = (MR^2 + I)\dot{\varphi}$$

$$+ \sum_{i=C,H} m_i \left(R^2\dot{\varphi} + (R - h)^2(\dot{\varphi} + \Omega_i) + 2R(R - h)(2\dot{\varphi} + \Omega_i)\cos\Omega_i t + \varphi \right)$$

$$\frac{\partial L}{\partial \varphi} = - \sum_{i=C,H} m_i(R - h)^2(\dot{\varphi} + \Omega_i)^2 + 2(R - h)R\dot{\varphi}(\dot{\varphi} + \Omega_i)\sin(\Omega_i t + \varphi)$$

$$- \sum_i m_i g(R - h)\sin(\Omega_i t + \varphi),$$

(A.13)

where $\sum_{i=C,H}$ denotes the sum of two cases, once when all the indices are equal to that of a cat $i = C$, and once when all indices are equal to that of a hamster $i = H$.

7. Looking for steady states, we need to set all-time derivatives to zero. Because of the explicit dependence of $\frac{\partial L}{\partial \varphi}$ on time, as is easy to observe from Eq. (A.13), there are no steady states.

8. To compute the energy E, we note that if L has a term linear in $\dot{\varphi}$, for example, $L_0 = A(\varphi, t)\dot{\varphi}$, then the corresponding term in energy $E_0 = \dot{\varphi}\frac{\partial L_0}{\partial \dot{\varphi}} - L_0$ vanishes exactly. So we don't have to compute the contribution of the linear terms to the energy E. The quadratic terms in $\dot{\varphi}$ in kinetic energy, on the other hand, remain unchanged. Thus, the energy E is computed as

$$E = \dot{\varphi}\frac{\partial L}{\partial \dot{\varphi}} - L =$$

$$= \frac{1}{2}\dot{\varphi}^2 \left[MR^2 + I + \sum_{i=C,H} m_i(R - h)^2 + 2R(R - h)\cos(\Omega_i t + \varphi) \right]$$

$$- \sum_{i=C,H} m_i g(R - h)\cos(\Omega_i t + \varphi).$$

(A.14)

9. Since the Lagrangian depends explicitly on time, the energy E in Eq. (A.14) is not conserved.

Problem 3 Unlike Problems 1 and 2, one variable is not enough to compute the complete state of the system since the location of the armadillo is not determined by the rotation of the circle. We thus need two variables, which could be taken as the rotation angle of the ball φ and the tilt angle of the vector connecting the armadillo's center with the center of the ball θ.

1. The coordinates of the ball, the hamster, and the center of the armadillo are as follows (if we take φ to be measured from the negative axis, so $\varphi = 0$ is down), so the configuration manifold is $Q = S^1 \times S^1$:

   (a) $\mathbf{x}_B = (R\varphi, 0)$.
   (b) $\mathbf{x}_H = (R\varphi + (R - h)\sin(\Omega_H t + \varphi), -(R - h)\cos(\Omega_H t + \varphi))$.
   (c) $\mathbf{x}_A = (R\varphi + (R - h)\sin(\theta), -(R - h)\cos(\theta))$.

   We have again assumed, for the simplicity of formulas, that the centers of mass of armadillo and the hamsters are the same distance h from the inner surface of the ball.

2. The velocities of the ball, the center of mass of the armadillo and the hamster are as follows:

   (a) $\dot{\mathbf{x}}_B = (R\dot\varphi, 0)$.
   (b) $\dot{\mathbf{x}}_H = (R\dot\varphi + (R - h)\cos(\Omega t + \varphi))(\Omega_H + \dot\varphi), (R - h)\sin(\Omega_H t + \varphi)(\Omega_H + \dot\varphi))$.
   (c) $\dot{\mathbf{x}}_A = (R\dot\varphi + (R - h)\dot\theta \cos\theta, (R - h)\dot\theta \sin(\theta))$.

3. To compute the kinetic energy, we need to remember the rotational kinetic energy of the armadillo. Suppose the absolute angle of the armadillo's rotation about its own axis is ψ. In that case, the condition that the armadillo rolls without slipping on the ball's surface is equivalent to the condition that the velocities of the contact point for the armadillo and the ball are equal. This condition is written as

$$h\dot\psi + (R - h)\dot\theta = R\dot\varphi, \quad \Rightarrow \quad \dot\psi = \frac{R}{h}\dot\varphi - \frac{(R - h)}{h}\dot\theta. \tag{A.15}$$

Then, the kinetic energy of the armadillo is composed of the rotational and translational energies, whereas the kinetic energies of the hamster and the ball remain the same. The computation of Lagrangian proceeds as follows:

(a) Kinetic energy of the ball:

$$T_B = \frac{1}{2}\left(M\dot{x}^2 + I\dot\varphi^2\right) = \frac{1}{2}\left(MR^2 + I\right)\dot\varphi^2.$$

(b) Kinetic energy of the hamster $T_H = \frac{1}{2}m\,|\dot{\mathbf{x}}_H|^2$ is computed as

$$T_H = \frac{1}{2}m_H\left[R^2\dot\varphi^2 + (R - h)^2(\dot\varphi + \Omega_H)^2\right.$$
$$\left. + 2(R - h)R\dot\varphi(\dot\varphi + \Omega_H)\cos(\Omega_H t + \varphi)\right].$$

(c) Kinetic energy of the armadillo $T_A = \frac{1}{2}m_A |\dot{\mathbf{x}}_A|^2 + \frac{1}{2}I_A\dot{\psi}^2$ is computed as

$$T_A = \frac{1}{2}m_A\left[R^2\dot{\varphi}^2 + (R-h)^2\dot{\theta}^2 + 2(R-h)R\dot{\varphi}\dot{\theta}\cos\theta\right]$$
$$+ \frac{1}{2h^2}I_A\left(R\dot{\varphi} - (R-h)\dot{\theta}\right)^2.$$

(d) Potential energy now relates to the hamster and the armadillo:

$$V = -m_Hg(R-h)\cos(\Omega_H t + \varphi) - m_Ag(R-h)\cos\theta.$$

(e) Lagrangian $L = T_B + T_H + T_A - V$:

$$L = \frac{1}{2}\left(MR^2 + I\right)\dot{\varphi}^2$$
$$+ \frac{1}{2}m_H\left[R^2\dot{\varphi}^2 + (R-h)^2(\dot{\varphi} + \Omega_H)^2\right.$$
$$+ 2(R-h)R\dot{\varphi}(\dot{\varphi} + \Omega_H)\cos(\Omega_H t + \varphi)]$$
$$+ \frac{1}{2}m_A\left[R^2\dot{\varphi}^2 + (R-h)^2\dot{\theta}^2 + 2(R-h)R\dot{\varphi}\dot{\theta}\cos\theta\right] \qquad \text{(A.16)}$$
$$+ \frac{1}{h^2}I_A\left(R\dot{\varphi} - (R-h)\dot{\theta}\right)^2$$
$$+ m_Hg(R-h)\cos(\Omega_H t + \varphi) + m_Ag(R-h)\cos\theta.$$

4. We see, again, that taking the location of the hamster $q = \varphi + \Omega_H t$ as one of the coordinates and leaving the other coordinate as before creates the Lagrangian that is time-independent:

$$L = \frac{1}{2}\left(MR^2 + I\right)(\dot{q} - \Omega)^2$$
$$+ \frac{1}{2}m_H\left[R^2(\dot{q} - \Omega)^2 + (R-h)^2\dot{q}^2 + 2(R-h)R(\dot{q} - \Omega)\dot{q}\cos q\right]$$
$$+ \frac{1}{2}m_A\left[R^2(\dot{q} - \Omega)^2 + (R-h)^2\dot{\theta}^2 + 2(R-h)R(\dot{q} - \Omega)\dot{\theta}\cos\theta\right]$$
$$+ \frac{1}{h^2}I_A\left(R(\dot{q} - \Omega) - (R-h)\dot{\theta}\right)^2$$
$$+ m_Hg(R-h)\cos q + m_Ag(R-h)\cos\theta.$$
$$\text{(A.17)}$$

5. Euler-Lagrange equations are given by

$$\frac{d}{dt}\frac{\partial L}{\partial \dot{q}} - \frac{\partial L}{\partial q} = 0$$

$$\frac{d}{dt}\frac{\partial L}{\partial \dot{\theta}} - \frac{\partial L}{\partial \theta} = 0, \quad \text{where we compute the derivatives as}$$

$$\frac{\partial L}{\partial \dot{q}} = (MR^2 + I)(\dot{q} - \Omega) + m_A \left[R^2(\dot{q} - \Omega) + 2(R - h)R\dot{\theta}\cos\theta \right]$$

$$+ m_H \left[R^2(\dot{q} - \Omega) + (R - h)^2\dot{q} + (R - h)R(2\dot{q} - \Omega)\cos q \right]$$

$$+ \frac{I_A}{h^2}R\left(R(\dot{q} - \Omega) - (R - h)\dot{\theta}\right)$$

$$\frac{\partial L}{\partial q} = -m_H R(R - h)\dot{q}(\dot{q} - \Omega)\sin q - m_H g(R - h)\sin q$$

$$\frac{\partial L}{\partial \dot{\theta}} = m_A \left[(R - h)^2\dot{\theta} + 2(R - h)R(\dot{q} - \Omega)\cos\theta \right]$$

$$+ \frac{I_A}{h^2}(R - h)\left((R - h)\dot{\theta} - R(\dot{q} - \Omega)\right)$$

$$\frac{\partial L}{\partial \theta} = -m_A(R - h)R(\dot{q} - \Omega)\dot{\theta}\sin\theta - m_A g(R - h)\sin\theta.$$

$$(A.18)$$

6. Critical points are determined by setting all the time derivatives to zero and also setting $\frac{\partial L}{\partial q} = \frac{\partial L}{\partial \theta} = 0$ in Eq. (A.18). These conditions give the critical points $\sin q = \sin\theta = 0$; that is, both the armadillo and the hamster are at the bottom. We use the energy condition to determine the stability.

7. The energy E is conserved since the Lagrangian (Eq. (A.17)) does not explicitly depend on time. To simplify the calculation, we remember that the terms in the Lagrangian that are linear in velocities $\dot{\theta}$ and $\dot{q}$ vanish in the expression for the energy. The term $L_1 = A(\theta, t)\dot{q}\dot{\theta}$ contributes to the energy as follows:

$$\dot{\theta}\frac{\partial L_1}{\partial \dot{\theta}} + \dot{q}\frac{\partial L_1}{\partial \dot{q}} - L_1 = A\dot{\theta}\dot{q} = L_1, \qquad (A.19)$$

so these terms are preserved, just as the quadratic terms proportional to $\dot{\theta}^2$ and $\dot{q}^2$, whereas the terms independent of $\dot{q}$ and $\dot{\theta}$ just change the sign. Thus, just looking at the expression for the Lagrangian (Eq. (A.17)), we write the energy E as

$$
\begin{aligned}
E = \dot{q}\frac{\partial L}{\partial \dot{q}} + \dot{\theta}\frac{\partial L}{\partial \dot{\theta}} - L = {} & \frac{1}{2}\left(MR^2 + I\right)\dot{q}^2 \\
& + \frac{1}{2}m_H\dot{q}^2\left[R^2 + (R-h)^2 + 2(R-h)R\cos q\right] \\
& + \frac{1}{2}m_A\left[R^2\dot{q}^2 + (R-h)^2\dot{\theta}^2 + 2(R-h)R\dot{q}\dot{\theta}\cos\theta\right] \\
& + \frac{1}{h^2}I_A\left(R\dot{q} - (R-h)\dot{\theta}\right)^2 \\
& - m_H g(R-h)\cos q - m_A g(R-h)\cos\theta + \text{const}
\end{aligned}
$$
(A.20)

(we can ignore the constant in energy).
8. At $(q = 0, \theta = 0)$ (both the hamster and armadillo down), the energy can be written as

$$
\begin{aligned}
E \simeq {} & \frac{1}{2}(MR^2 + I)\dot{q}^2 + \frac{1}{2}m_H\dot{q}^2(2R - h)^2 + \frac{1}{h^2}I_A\left(R\dot{q} - (R-h)\dot{\theta}\right)^2 \\
& + \frac{1}{2}m_A\left(R\dot{q} + (R-h)\dot{\theta})^2\right) + \frac{1}{2}m_A g(R-h)q^2 + \frac{1}{2}m_A g(R-h)\theta^2
\end{aligned}
$$
(A.21)

Thus, locally, the energy is a sum of squares, and the critical point is stable. For $q = \theta = \pi$, or one of these variables being π, the sign in the approximation of $\cos q$ and/or $\cos\theta$ changes sign. Energy is no longer positive, definite, and in a quadratic form. It does not necessarily indicate the instability, but the instability can be rigorously shown by linearization, which we are not going to do here.

Problem 4

1. If we take $E_0 = \frac{1}{2}\dot{q}^2 + U(q)$, then

$$
\frac{dE_0}{dt} = \left(\ddot{q} + U'(q)\right)\dot{q} = -\gamma\dot{q}^2 \leq 0.
$$
(A.22)

2. We are looking for Lagrangians $L(q, \dot{q}, t)$ such that the Euler-Lagrange equations are equivalent to the equations of the motion of the oscillator (Eq. (3.50)) multiplied by some function $\mu(t)$:

$$
\frac{d}{dt}\frac{\partial L}{\partial \dot{q}} - \frac{\partial L}{\partial q} = 0 \quad \Leftrightarrow \quad \mu(t)\ddot{q} + \gamma\mu(t)\dot{q} + \mu(t)U'(q) = 0
$$
(A.23)

Then, it is reasonable to look for $\mu(t)$ such that $\dot{\mu} = \gamma\mu$, or $\mu = e^{\gamma t}$. Then, Euler-Lagrange equations (Eq. (A.23)) are consistent for

$$L(q, \dot{q}, t) = e^{\gamma t}\left(\frac{1}{2}\dot{q}^2 - U(q)\right). \tag{A.24}$$

3. The energy E for this Lagrangian

$$E = \dot{q}\frac{\partial L}{\partial \dot{q}} - L = e^{\gamma t}\left(\frac{1}{2}\dot{q}^2 + U(q)\right) = e^{\gamma t}E_0. \tag{A.25}$$

4. Since L explicitly depends on time, we get

$$\frac{dE}{dt} = -\frac{\partial L}{\partial t} = -\gamma e^{\gamma t}\left(\frac{1}{2}\dot{q}^2 - U(q)\right). \tag{A.26}$$

On the other hand, since $E = e^{\gamma t}E_0$, Eq. (A.26) gives

$$\frac{dE}{dt} = \gamma e^{\gamma t}E_0 + e^{\gamma t}\frac{dE_0}{dt} = -\gamma e^{\gamma t}\left(\frac{1}{2}\dot{q}^2 - U(q)\right), \quad \Rightarrow \frac{dE_0}{dt} = -\gamma\dot{q}^2, \tag{A.27}$$

consistent with Eq. (A.22).

Problem 5 Consider the elastic pendulum with the configuration manifold $Q = S^1 \times \mathbb{R}$. We could also take $Q = \mathbb{R}^2$, of course, but the computations get a bit more complex.

1. Coordinates $\mathbf{x} = l(t)(\sin\varphi(t), -\cos\varphi(t))$.
2. Velocities $\mathbf{v} = \dot{\mathbf{x}} = \dot{l}(t)(\sin\varphi, -\cos\varphi) + l\dot{\varphi}(\cos\varphi, \sin\varphi)$.
3. Compute the derivatives of velocities with respect to $(\dot{l}, \dot{\varphi})$ and verify the cancellation of dots formula:

$$\frac{\partial\mathbf{v}}{\partial\dot{l}} = (\sin\varphi, -\cos\varphi) = \frac{\partial\mathbf{x}}{\partial l}$$

$$\frac{\partial\mathbf{v}}{\partial\dot{\varphi}} = l(\cos\varphi, \sin\varphi) = \frac{\partial\mathbf{x}}{\partial\varphi}. \tag{A.28}$$

4. We can compute Rayleigh's dissipation function $\mathcal{R} = \frac{1}{2}\alpha|\mathbf{v}|^2$ as

$$\mathcal{R} = \frac{\alpha}{2}\left(\dot{l}^2 + l^2\dot{\varphi}^2\right), \quad F_l = -\frac{\partial\mathcal{R}}{\partial\dot{l}} = -\alpha\dot{l}, \quad F_\varphi = -\frac{\partial\mathcal{R}}{\partial\dot{\varphi}} = -\alpha l^2\dot{\varphi}. \tag{A.29}$$

5. The generalized forces in l and φ can also be computed as

$$F_l = \mathbf{F} \cdot \frac{\partial \mathbf{v}}{\partial \dot{l}} = -\alpha \mathbf{v} \cdot \frac{\partial \mathbf{v}}{\partial \dot{l}} = -\alpha \dot{l}$$

$$F_\varphi = \mathbf{F} \cdot \frac{\partial \mathbf{v}}{\partial \dot{\varphi}} = -\alpha \mathbf{v} \cdot \frac{\partial \mathbf{v}}{\partial \dot{\varphi}} = -\alpha l^2 \dot{\varphi}.$$

$$(\text{A.30})$$

6. The Lagrangian and the dissipative Euler-Lagrange equations are then

$$L = \frac{1}{2} m |\mathbf{v}|^2 - U(l)$$

$$\frac{d}{dt}\frac{\partial L}{\partial \dot{\varphi}} - \frac{\partial L}{\partial \varphi} = -\alpha \dot{l} \qquad\qquad (\text{A.31})$$

$$\frac{d}{dt}\frac{\partial L}{\partial \dot{l}} - \frac{\partial L}{\partial l} = -\alpha l^2 \dot{\varphi}.$$

7. The energy

$$E = \dot{\varphi}\frac{\partial L}{\partial \dot{\varphi}} + \dot{l}\frac{\partial L}{\partial \dot{l}} - L = \frac{1}{2} m |\mathbf{v}|^2 + U(l). \qquad (\text{A.32})$$

8. Evolution of energy is computed as

$$\frac{dE}{dt} = \dot{\varphi}F_\varphi + \dot{l}F_l = -\alpha \left(\dot{l}^2 + l^2 \dot{\varphi}^2 \right) = -2\mathcal{R} \le 0, \qquad (\text{A.33})$$

so the system is dissipative.

Problem 6 The configuration manifold of this problem is $Q = S^1$. If φ is the angle of the bead with respect to the vertical in the plane of the circle, the evolution angle about the axis is ωt.

1. The coordinates in $\mathbb{R}^3$ are computed as $\mathbf{x} = R(\sin \varphi \cos \omega t, \sin \varphi \sin \omega t, -\cos \varphi)$.
2. The velocities are

$$\dot{\mathbf{x}} = \mathbf{v} = R\dot{\varphi}(\cos \varphi \cos \omega t, \cos \varphi \sin \omega t, \sin \varphi)$$

$$+ R\omega(-\sin \varphi \sin \omega t, \sin \varphi \cos \omega t, 0).$$

3. Partial derivatives (verify cancellation of dots!)

$$\frac{\partial \mathbf{v}}{\partial \dot{\varphi}} = \frac{\partial \mathbf{x}}{\partial \varphi} = R(\cos \varphi \cos \omega t, \cos \varphi \sin \omega t, \sin \varphi). \qquad (\text{A.34})$$

4. The actual force is $\mathbf{F} = -\alpha \mathbf{v}$. Rayleigh's dissipation function can be computed as

$$\mathcal{R} = \frac{1}{2}\alpha |\mathbf{v}|^2 = \frac{1}{2}\left(R^2\dot{\varphi}^2 + R^2\omega^2 \sin^2\varphi\right). \tag{A.35}$$

The generalized friction force can be computed either as simply

$$F_\varphi = -\frac{\partial \mathcal{R}}{\partial \dot{\varphi}} = -\alpha R^2 \dot{\varphi}$$

or, alternatively, through the work done by generalized forces:

$$F_\varphi = \mathbf{F} \cdot \frac{\partial \mathbf{v}}{\partial \dot{\varphi}} = -\alpha \mathbf{v} \cdot \frac{\partial \mathbf{v}}{\partial \dot{\varphi}} = -\alpha R^2 \dot{\varphi},$$

which gives the same result.

5. For the potential energy $U(\varphi) = -mgR\cos\varphi$, the Lagrangian is

$$L = \frac{1}{2}m|\mathbf{v}|^2 - U(\varphi) = \frac{1}{2}mR^2\left(\dot{\varphi}^2 + \omega^2\sin^2\varphi\right) + mgR\cos\varphi.$$

We can define the "effective potential energy" $U_e = U(q) - \frac{1}{2}mR^2\omega^2\sin^2\varphi$. Note that if ω is sufficiently large, $\varphi = 0$ becomes unstable since the negative ("unstable") part of the effective energy $-\frac{1}{2}mR^2\omega^2\sin^2\varphi$ dominates over the positive part $U(\varphi) = -mgR\cos\varphi$. The precise condition of the instability of the bottom equilibrium $\varphi = 0$ is $\omega > \Omega$, where $\Omega = \sqrt{g/R}$.

6. Euler-Lagrange equations:

$$\frac{d}{dt}\frac{\partial L}{\partial \dot{\varphi}} - \frac{\partial L}{\partial \varphi} = F_\varphi \Rightarrow mR^2\ddot{\varphi} - mR^2\omega^2\sin\varphi\cos\varphi + mgR\sin\varphi = -\alpha R^2\dot{\varphi} \tag{A.36}$$

or, alternatively,

$$\ddot{\varphi} + \sin\varphi\left(\Omega^2 - \omega^2\cos\varphi\right) = -\frac{\alpha}{m}\dot{\varphi}, \quad \Omega := \sqrt{\frac{g}{R}}.$$

7. The energy is

$$E = \dot{\varphi}\frac{\partial L}{\partial \dot{\varphi}} - L = \frac{1}{2}mR^2\left(\dot{\varphi}^2 - \omega^2\sin^2\varphi\right) - mgR\cos\varphi. \tag{A.37}$$

Note that this energy is *not* the kinetic plus potential energy since the part of the kinetic energy proportional to $\frac{1}{2}mR^2\omega^2 \sin^2\varphi$ enters with a minus sign:

$$\frac{dE}{dt} = F_\varphi\dot\varphi = -\alpha R^2\dot\varphi^2 \leq 0$$

If, instead of the energy (Eq. (A.37)), we took the kinetic plus potential energy, then we would no longer have $\dot E \leq 0$, and the system would not be dissipative.

Note that $\dot E \neq -2\mathcal{R}$ for Rayleigh's dissipation function given by Eq. (A.35); however, if we define another Rayleigh's dissipation function as the "dissipation potential," for example, $\widetilde{\mathcal{R}} = \frac{1}{2}\alpha R^2\dot\varphi^2$, then $F_\varphi = -\frac{\partial\widetilde{\mathcal{R}}}{\partial\dot\varphi}$ and $\dot E = -2\widetilde{\mathcal{R}}$. Thus, if one uses Rayleigh's function definition as simply the "dissipation potential," and not as $\mathcal{R} = \frac{1}{2}\alpha|\mathbf{v}|^2$, then one could in principle define it in such a way that $\dot E = -2\mathcal{R}$.

A.4 Chapter 4

Problem 1 Let us solve this problem for a more general case:

$$L = \dot q^1\dot q^2 - U(\xi), \xi := q^1q^2.$$

The answer is going to be exactly the same. The Lagrangian is invariant with respect to the transformation $q^1 \to q^1 e^\epsilon, q^1 \to q^2 e^{-\epsilon}$. Then, we obtain

$$\delta\mathbf{q} = \begin{pmatrix} q^1 \\ -q^2 \end{pmatrix}, \quad N = \frac{\partial L}{\partial\dot{\mathbf{q}}}\cdot\delta\mathbf{q} = \dot q^2q^1 - \dot q^1q^2.$$

Problem 2 Suppose $L = \frac{1}{2}\left(\dot q_1^2 + \dot q_2^2\right) - U(q_1, q_2)$. Let us first solve this problem without assuming that Γ comes from symmetry and Noether's theorem. The Euler-Lagrange equations are simply $\ddot q_i = -\frac{\partial U}{\partial q_i}$, so differentiating Γ with respect to time gives

$$\Gamma = q_1\dot q_2 - q_2\dot q_1, \quad \dot\Gamma = q_1\ddot q_2 - q_2\ddot q_1 = -q_1\frac{\partial U}{\partial q_2} + q_2\frac{\partial U}{\partial q_1} = 0. \qquad (A.38)$$

Equation (A.43) is an equation for U. One way to solve this equation is to use the method of characteristics. The equations for characteristics are

$$\frac{dq_1}{q_2} = -\frac{dq_2}{q_1}, \quad \Rightarrow \quad q_1^2 + q_2^2 = C \quad \Rightarrow U = U(r), \quad r^2 := q_1^2 + q_2^2. \qquad (A.39)$$

If you don't want to use the equation of characteristics, you can solve this problem in a straightforward way as follows. Instead of $U = U(q_1, q_2)$, we write $U = U(r, \varphi)$

(somewhat abusing the notation) in polar coordinates, so

$$
\begin{pmatrix} q_1 \\ q_2 \end{pmatrix} = \begin{pmatrix} r\cos\varphi \\ r\sin\varphi \end{pmatrix}, \quad \begin{pmatrix} r \\ \varphi \end{pmatrix} = \begin{pmatrix} \sqrt{q_1^2 + q_2^2} \\ \arctan\frac{q_2}{q_1} \end{pmatrix}, \tag{A.40}
$$

and the derivatives can be computed as

$$
\begin{aligned}
\frac{\partial U}{\partial q_1} &= \frac{\partial r}{\partial q_1}\frac{\partial U}{\partial r} + \frac{\partial\varphi}{\partial q_1}\frac{\partial U}{\partial\varphi} = \frac{q_1}{r}\frac{\partial U}{\partial r} - \frac{q_2}{r^2}\frac{\partial U}{\partial\varphi} \\
\frac{\partial U}{\partial q_2} &= \frac{\partial r}{\partial q_2}\frac{\partial U}{\partial r} + \frac{\partial\varphi}{\partial q_2}\frac{\partial U}{\partial\varphi} = \frac{q_2}{r}\frac{\partial U}{\partial r} + \frac{q_1}{r^2}\frac{\partial U}{\partial\varphi}.
\end{aligned} \tag{A.41}
$$

Then, the condition $\delta U = 0$ given by Eq. (A.43) gives simply $\frac{\partial U}{\partial\varphi} = 0$, so $U = U(r)$ in agreement with Eq. (A.39).

If we assume that Γ comes from symmetry, then it can be expressed in Noether's integral form:

$$
\Gamma = q_1\dot{q}_2 - q_2\dot{q}_1 = \frac{\partial L}{\partial\dot{q}_1}\delta q_1 + \frac{\partial L}{\partial\dot{q}_2}\delta q_2 = \dot{q}_1\delta q_1 + \dot{q}_2\delta q_2. \tag{A.42}
$$

If we say that Γ is Noether's integral, we obtain $\delta q_1 = -q_2$, $\delta q_2 = q_1$. This is the infinitesimal rotation symmetry. The kinetic energy is invariant with respect to that symmetry, so for the invariance of Lagrangian, we must have

$$
\delta U = \frac{\partial U}{\partial q_1}\delta q_1 + \frac{\partial U}{\partial q_2}\delta q_2 = -\frac{\partial U}{\partial q_1}q_2 + \frac{\partial U}{\partial q_2}q_1 = 0, \tag{A.43}
$$

leading exactly to the PDE for $U(q_1, q_2)$ derived in Eq. (A.38). Thus, the answer is $U = U(r)$ with $r = \sqrt{q_1^2 + q_2^2}$.

Problem 3

1. The Lagrangian is $L = \frac{1}{2}\frac{|\dot{\mathbf{q}}|^2}{|\mathbf{q}|^2}$, and the Euler-Lagrange equations are then

$$
\frac{d}{dt}\frac{\dot{\mathbf{q}}}{|\mathbf{q}|^2} + \frac{|\dot{\mathbf{q}}|^2}{|\mathbf{q}|^4}\mathbf{q} = 0 \tag{A.44}
$$

2. Clearly, substitution of $\mathbf{q} \to e^\epsilon \mathbf{q}$ multiplies both $\mathbf{q}$ and $\dot{\mathbf{q}}$ by the same coefficient and leaves the Lagrangian invariant. Since $\delta\mathbf{q} = \mathbf{q}$, we obtain

$$
N_1 = \frac{\partial L}{\partial\dot{\mathbf{q}}}\cdot\delta\mathbf{q} = \frac{\dot{\mathbf{q}}\cdot\mathbf{q}}{|\mathbf{q}|^2} \tag{A.45}
$$

Interestingly, because of the functional shape of N_1 given by Eq. (A.45), we can reach additional conclusion about $|\mathbf{q}|$ since

$$N_1 = \frac{d}{dt} \log |\mathbf{q}|^2 = \text{const}, \qquad |\mathbf{q}|^2 = Ce^{N_1 t}. \tag{A.46}$$

Using Eq. (A.46), the Euler-Lagrange equation can be simplified somewhat, which we won't do here.

3. For the rotations about the axis $\mathbf{n}$, we have seen $\delta\mathbf{q} = -\mathbf{n} \times \mathbf{q}$. Noether's integral is then

$$N_2 = \frac{\partial L}{\partial \dot{\mathbf{q}}} \cdot \delta\mathbf{q} = -\frac{\partial L}{\partial \dot{\mathbf{q}}} \cdot (\mathbf{n} \times \mathbf{q}) = \frac{\dot{\mathbf{q}} \times \mathbf{q}}{|\mathbf{q}|^2} \cdot \mathbf{n}. \tag{A.47}$$

Since $\mathbf{n}$ is arbitrary, we have a conservation of the vector quantity:

$$N_2 = \mathbf{q} \times \mathbf{p}, \qquad \mathbf{p} := \frac{\partial L}{\partial \dot{\mathbf{q}}} = \frac{\dot{\mathbf{q}}}{|\mathbf{q}|^2}, \tag{A.48}$$

which is in fact the conservation of angular momentum. One can also rewrite Eq. (A.48) using Eq. (A.46) as

$$\mathbf{q} \times \dot{\mathbf{q}} = N_2 e^{N_1 t}, \tag{A.49}$$

which is a time-dependent (classical) angular momentum conservation law.

A.5   Chapter 5

Problem 1 Let us introduce rescaling of all variables in terms of l, and define the initial position vectors of the support as $\mathbf{x}_i^S = l\mathbf{V}_i$, where $|\mathbf{V}_i| = 1$ and the points are enumerated counterclockwise:

$$V_1 = (1, 0), \quad V_2 = (0, 1), \quad V_3 = (-1, 0), \quad V_4 = (0, -1). \tag{A.50}$$

The configuration manifold of the particle is $\mathbb{R}^2$. If the absolute coordinate of the mass is $\mathbf{x}$, let us also introduce the scaled variables $\mathbf{X} = \mathbf{x}/l$, with $|\mathbf{X}| \ll 1$. The length of each spring i is then computed as

$$l_i = l\,|\mathbf{V}_i + \mathbf{X}_i| = l\sqrt{\mathbf{V}_i \cdot \mathbf{V}_i + 2\mathbf{V}_i \cdot \mathbf{X} + \mathbf{X} \cdot \mathbf{V}} \simeq 1 + \mathbf{V}_i \cdot \mathbf{X} + \dots.$$

Since the potential energy is going to be a square of the spring's extension $l_i - l$, we will only need to compute the lowest-order term:

$$U_i = \frac{c_i}{2}(l_i - l)^2 \simeq \frac{c_i}{2}l^2 (\mathbf{V}_i \cdot \mathbf{X})^2$$

Using (A.50), and remembering that $c_1 = c_3 = \mu c$ and $c_2 = c_4 = c$, we arrive to the total potential energy:

$$U = \mu c l^2 x_1^2 + c l^2 x_2^2, \quad \Rightarrow \quad \mathbb{K} = 2c \begin{pmatrix} \mu & 0 \\ 0 & 1 \end{pmatrix}. \tag{A.51}$$

The mass matrix is easy to compute from the kinetic energy:

$$T = \frac{1}{2}ml^2|\dot{\mathbf{X}}^2| = \frac{1}{2}ml^2(\dot{X}_1^2 + \dot{X}_2^2), \quad \Rightarrow \quad \mathbb{M} = ml^2 \begin{pmatrix} 1 & 0 \\ 0 & 1 \end{pmatrix}. \tag{A.52}$$

The matrix $\Lambda = \mathbb{M}^{-1}\mathbb{K}$ is a diagonal matrix:

$$\Lambda = \omega_0^2 \begin{pmatrix} \mu & 0 \\ 0 & 1 \end{pmatrix}. \tag{A.53}$$

Since the matrix is diagonal, the eigenvalues are $\lambda_1 = \omega_0^2\mu$ and $\lambda_2 = \omega_0^2$ with the eigenvectors $\mathbf{v}_1 = (1,0)^T$ and $\mathbf{v}_2 = (0,1)^T$. The general solution for small oscillation is

$$\mathbf{x} = (A \cos \sqrt{\mu}\omega_0 t + B \sin \sqrt{\mu}\omega_0 t) \begin{pmatrix} 1 \\ 0 \end{pmatrix} + (C \cos \omega_0 t + D \sin \omega_0 t) \begin{pmatrix} 0 \\ 1 \end{pmatrix}$$

$$= \begin{pmatrix} A \cos \sqrt{\mu}\omega_0 t + B \sin \sqrt{\mu}\omega_0 t \\ C \cos \omega_0 t + D \sin \omega_0 t \end{pmatrix}.$$

$$\tag{A.54}$$

Problem 2

1. Configuration manifold of the problem is $Q = S^1 \times S^1$. The coordinates of each bob are

$$\mathbf{x}_1 = (l \sin \varphi_1, -l \cos \varphi_1), \quad \mathbf{x}_2 = (L + l \sin \varphi_2, -l \cos \varphi_2).$$

2. Velocities and the kinetic energy are given by

$$\dot{\mathbf{x}}_1 = (l \cos \varphi_1, l \sin \varphi_1)\dot{\varphi}_1, \quad \dot{\mathbf{x}}_2 = (l \cos \varphi_2, l \sin \varphi_2)\dot{\varphi}_2, \quad T = \frac{1}{2}ml^2 \left(\dot{\varphi}_1^2 + \dot{\varphi}_2^2 \right).$$

3. The gravitational potential energy is $U_g = -mgl(\cos\varphi_1 + \cos\varphi_2)$. To compute the elastic energy, we find the spring extension as

$$|\mathbf{x}_1 - \mathbf{x}_2|$$

$$= \sqrt{L^2 + 2Ll(\sin\varphi_1 - \sin\varphi_2) + l^2\left((\sin\varphi_1 - \sin\varphi_2)^2 + (\cos\varphi_1 - \cos\varphi_2)^2\right)}.$$

We again only need to find the expression of $|\mathbf{x}_1 - \mathbf{x}_2| - L$ that is linear in (φ_1, φ_2). Expanding the expression above under the assumption of small angles to the linear approximation in $\varphi_1 - \varphi_2$, we obtain

$$d = |\mathbf{x}_1 - \mathbf{x}_2| - L \simeq l(\varphi_1 - \varphi_2).$$

The elastic potential energy is $U_e = \frac{1}{2}cd^2$, and the total potential energy, truncated to the second order in angles, gives

$$U = U_e + U_g \simeq \frac{1}{2}cl^2(\varphi_1 - \varphi_2)^2 + \frac{1}{2}mgl(\varphi_1^2 + \varphi_2^2).$$

4. Defining two eigenfrequencies $\omega_0^2 = g/l$ (the gravitation frequency) and $\omega_1^2 = c/m$ (the elastic frequency), and the ratio of the square frequencies $\alpha = \omega_1^2/\omega_0^2$, the matrices $\mathbb{K}$ and $\Lambda = \mathbb{M}^{-1}\mathbb{K}$ are given by

$$\mathbb{K} = mgl\begin{pmatrix} 1+\alpha & -\alpha \\ -\alpha & 1+\alpha \end{pmatrix}, \quad \Lambda = \omega_0^2\begin{pmatrix} 1+\alpha & -\alpha \\ -\alpha & 1+\alpha \end{pmatrix}.$$

5. Characteristic polynomial for Λ/ω_0^2 is given by $(1+\alpha-\lambda)^2-\alpha^2$, so $\lambda = 1+\alpha\pm\alpha$, or $\lambda_1 = 1 + 2\alpha$ and $\lambda_2 = 1$. The corresponding eigenvectors are $\mathbf{v}_1 = (1, -1)^T$ and $\mathbf{v}_2 = (1, 1)^T$. Then, we can define $\Omega_1 = \sqrt{1 + 2\alpha}\omega_0 = \sqrt{2\omega_1^2 + \omega_0^2}$, and the solutions for small-scale oscillations are given by

$$\begin{pmatrix} \varphi_1 \\ \varphi_2 \end{pmatrix} = (A\cos\Omega_1 t + B\sin\Omega_1 t)\begin{pmatrix} 1 \\ -1 \end{pmatrix} + (C\cos\omega_0 t + D\sin\omega_0 t)\begin{pmatrix} 1 \\ 1 \end{pmatrix}.$$

Problem 3

1. We parameterize the n-th ball by the deviation of the angle from its equilibrium position, φ_n. The extension of the spring for particles at n and $n+1$ is $l(\varphi_{n+1} - \varphi_n)$. The total potential energy is

$$U = \frac{c}{2}\sum_{n=1}^{N}(\varphi_{n+1} - \varphi_n)^2, \quad \varphi_{N+1} = \varphi_1.$$

2. The absolute velocity of the n-th particle is $l\dot{\varphi}_n$, so the kinetic energy is $T = \frac{1}{2}ml^2 \sum_n \dot{\varphi}_n^2$. The mass matrix is $\mathbb{M} = ml^2\mathbb{I}_n$ (the n-dimensional unity matrix). The matrices $\mathbb{K}$ and $\Lambda = \mathbb{M}^{-1}\mathbb{K}$ are given by

$$
\mathbb{K} = cl^2\mathbb{D}_2\,, \quad \Lambda = \omega_0^2\mathbb{D}_2\,, \quad \mathbb{D}_2 := \begin{pmatrix} 2 & -1 & \dots & 0 & -1 \\ -1 & 2 & -1 & \dots & 0 \\ & & \dots & & \\ -1 & 0 & \dots & -1 & 2 \end{pmatrix}, \qquad (A.55)
$$

where $\omega_0 = \sqrt{c/m}$, $\mathbb{I}_n$ is the n-dimensional unity matrix and $\mathbb{D}_2$ is the discrete analog of the two-dimensional derivative.

3. To find eigenvectors/eigenvalues of the matrix Λ, we first find eigenvalues of the dimensionless matrix $\mathbb{D}_2$. Consider $\mathbf{v}_q = (1, q, q^2, \dots, q^{N-1})$ for some complex number q. Multiplying Λ described by Eq. (A.55), we see that $\mathbb{D}_2\mathbf{v}_q$ has every component between 2 and $N - 1$ is multiplied by the factor $\lambda = (2 - q - q^{-1})$, except for the first and the last component. The first component of $\Lambda\mathbf{v}_q$ is multiplied by $\lambda_0 = (2 - q - q^{N-1})$ and the last component is multiplied by $\lambda_N = (2 - q^{-1} - q^{-(N-1)})$. Then, for $\mathbf{v}_q$ to be an eigenvector, we must have $\lambda_0 = \lambda_N = \lambda$, which is only possible when $q^{N-1} = q^{-1}$, or $q^N = 1$.

   The equation $q^N = 1$ has exactly N solutions in the complex plane: the N-th roots of unity. All these roots have magnitude 1 and are described by the formula $q_m = e^{i2\pi m/N}$, $m = 0, \dots, N - 1$. Since there are exactly N eigenvalues and N eigenvectors, we have found all eigenvalues and eigenvectors of the matrix $\mathbb{D}_2$. The eigenvalues of Λ are then obtained as the corresponding eigenvalues of $\mathbb{D}_2$ multiplied by ω_0^2.

4. It is a good idea to make the eigenvalues and eigenvectors real. In fact, the eigenvalues of $\mathbb{D}_2$, which we denote λ_m, are already real since

$$
\begin{aligned}
\lambda_m &= 2 - q_m - q_m^{-1} = 2 - e^{i2\pi m/N} - e^{-i2\pi m/N} \\
&= 2 - 2\cos\left(2\pi\frac{m}{N}\right) = 4\sin^2\left(\pi\frac{m}{N}\right).
\end{aligned} \qquad (A.56)
$$

The eigenfrequencies are then $\omega_m = \omega_0\sqrt{\lambda_m} = 2\sin\left(\pi\frac{m}{N}\right)\omega_0$. One can take the real and imaginary parts of the eigenvectors as the new eigenvectors. Note that $m = 0$ corresponds to $\omega = 0$. Physically, that zero eigenfrequency corresponds to all masses shifting along the circle with the same speed.

A.6   Chapter 6

Problem 1

(a) The Lagrangian is

$$L = \frac{|\dot{\mathbf{q}}|^2}{2|\mathbf{q}|^2} - U(\mathbf{q});$$

then

$$\mathbf{p} = \frac{\partial L}{\partial \dot{\mathbf{q}}} = \frac{\mathbf{q}}{|\mathbf{q}|^2}, \quad \dot{\mathbf{q}}(\mathbf{q}, \mathbf{p}) = |\mathbf{q}|^2 \mathbf{p}.$$

So the Hamiltonian is

$$H = \mathbf{p} \cdot \dot{\mathbf{q}}(\mathbf{p}, \mathbf{q}) - L(\mathbf{q}, \dot{\mathbf{q}}(\mathbf{p}, \mathbf{q})) = \frac{1}{2}|\mathbf{q}|^2 |\mathbf{p}|^2 + U(\mathbf{q}).$$

Hamilton's equations are thus

$$\dot{\mathbf{p}} = -\frac{\partial H}{\partial \mathbf{q}} = -|\mathbf{p}|^2 \mathbf{q} - \frac{\partial U}{\partial \mathbf{q}}, \quad \dot{\mathbf{q}} = \frac{\partial H}{\partial \mathbf{p}} = |\mathbf{q}|^2 \mathbf{p}.$$

(b) The Lagrangian is

$$L = \frac{1}{2}\dot{\mathbf{q}}^T \mathbb{A}\dot{\mathbf{q}} + \mathbf{F} \cdot \dot{\mathbf{q}},$$

so the momenta $\mathbf{p}(\dot{\mathbf{q}}, \mathbf{q})$ in terms of velocities, and velocities in terms of momenta $\dot{\mathbf{q}}(\mathbf{p}, \mathbf{q})$, are computed as

$$\mathbf{p} = \frac{\partial L}{\partial \mathbf{q}} = \mathbb{A}\dot{\mathbf{q}} + \mathbf{F}, \quad \dot{\mathbf{q}} = \mathbb{A}^{-1}(\mathbf{p} - \mathbf{F}).$$

Thus, the Hamiltonian is

$$H(\mathbf{p}, \mathbf{q}) = \dot{\mathbf{q}} \cdot \mathbf{p} - L = \mathbf{p} \cdot \mathbb{A}^{-1}(\mathbf{p} - \mathbf{F}) -$$

$$\left(\frac{1}{2}(\mathbf{p} - \mathbf{F})^T \left(\mathbb{A}^{-1}\right)^T \mathbb{A} \left(\mathbb{A}^{-1}(\mathbf{p} - \mathbf{F})\right) + \mathbf{F} \cdot \mathbb{A}^{-1}(\mathbf{p} - \mathbf{F}) \right).$$

$$(A.57)$$

However, remember that $\mathbb{A}^T = \mathbb{A}$, so $(\mathbb{A}^{-1})^T = \mathbb{A}^{-1}$. With that, Eq. (A.57) is computed as

$$H(\mathbf{p}, \mathbf{q}) = \frac{1}{2}\mathbf{p}^T \mathbb{A}^{-1}\mathbf{p} - \mathbf{p}^T \mathbb{A}^{-1}\mathbf{F} + \frac{1}{2}\mathbf{F}^T \mathbb{A}^{-1}\mathbf{F}. \tag{A.58}$$

Defining $\mathbb{B}(\mathbf{q}) = \mathbb{A}^{-1}(\mathbf{q})$, we write Hamilton's equations as

$$\dot{\mathbf{q}} = \frac{\partial H}{\partial \mathbf{p}} = \mathbb{A}^{-1}(\mathbf{p} - \mathbf{F}) = \mathbb{B}(\mathbf{p} - \mathbf{F})$$

$$\dot{\mathbf{p}} = -\frac{\partial H}{\partial \mathbf{q}} = \frac{1}{2}\mathbf{p}^T \frac{\partial \mathbb{B}}{\partial \mathbf{q}}\mathbf{p} - \mathbf{p}^T \frac{\partial}{\partial \mathbf{q}}(\mathbb{B}(\mathbf{q})\mathbf{F}(\mathbf{q})) + \frac{1}{2}\frac{\partial}{\partial \mathbf{q}}\left(\mathbf{F}^T \mathbb{B}\mathbf{F}\right). \tag{A.59}$$

Problem 2, part (a) Configuration manifold is $\mathbb{R} \times S^1$. The coordinates and velocities of masses are

$$\mathbf{x}_M = (x, 0), \quad \mathbf{x}_m = (x, 0) + l(\sin\varphi, -\cos\varphi),$$

$$\mathbf{v}_M = (\dot{x}, 0), \quad \mathbf{v}_m = (\dot{x}, 0) + l\dot{\varphi}(\cos\varphi, \sin\varphi).$$

The Lagrangian is then

$$L = T - U = \frac{1}{2}M|\mathbf{v}_M|^2 + \frac{1}{2}m|\mathbf{v}_m|^2 + mgl\cos\varphi$$

$$= \frac{1}{2}(M + m)\dot{x}^2 + ml\dot{x}\dot{\varphi}\cos\varphi + \frac{1}{2}ml^2\dot{\varphi}^2 + mgl\cos\varphi.$$

The momenta are given by

$$\begin{cases} p_x = \dfrac{\partial L}{\partial \dot{x}} = (M + m)\dot{x} + ml\dot{\varphi}\cos\varphi \\[2mm] p_\varphi = \dfrac{\partial L}{\partial \dot{\varphi}} = ml\dot{x}\cos\varphi + ml^2\dot{\varphi} \end{cases} \Rightarrow \mathbf{p} = \mathbb{M}\mathbf{q}, \quad \mathbb{M} = \begin{pmatrix} (M + m) & ml\cos\varphi \\ ml\cos\varphi & ml^2 \end{pmatrix}.$$

The inverse of the mass matrix M is given by

$$\mathbb{M}^{-1} = \frac{1}{D(\varphi)}\begin{pmatrix} ml^2 & -ml\cos\varphi \\ -ml\cos\varphi & M + m \end{pmatrix}, \quad D(\varphi) := \det\mathbb{M} = Mml^2 + m^2l^2\sin^2\varphi.$$

Then, since the Lagrangian is quadratic in velocities, we can use Eq. (6.9) with $U(\varphi) = -mgl \cos \varphi$ to write the Hamiltonian:

$$H = \frac{1}{2}\mathbf{p}^T \mathbb{M}^{-1}\mathbf{p} + U(\mathbf{q}) = \mathbf{p}^T \frac{1}{2D}\left(ml^2 p_x^2 - 2ml^2 \cos^2 \varphi p_x p_\varphi + (M+m)p_\varphi^2\right)$$
$$- mgl \cos \varphi$$

with $D(\varphi) = \det\mathbb{M} = Mml^2 + m^2l^2 \sin^2 \varphi$. Hamilton's equations are then given by

$$\dot{x} = \frac{\partial H}{\partial p_x} = \frac{1}{D(\varphi)}\left(ml^2 p_x - p_\varphi ml^2 \cos^2 \varphi\right)$$

$$\dot{\varphi} = \frac{\partial H}{\partial p_\varphi} = \frac{1}{D(\varphi)}\left(-ml^2 p_x \cos^2 \varphi + (M+m)p_\varphi\right)$$

$$\dot{p}_x = -\frac{\partial H}{\partial x} = 0$$

$$\dot{p}_\varphi = -\frac{\partial H}{\partial \varphi} = \frac{ml^2 \sin \varphi \cos \varphi}{D(\varphi)^2}\left(ml^2 p_x^2 - 2ml^2 \cos^2 \varphi p_x p_\varphi + (M+m)p_\varphi^2\right)$$
$$- \frac{1}{D(\varphi)}\left(ml^2 p_x p_\varphi \sin \varphi \cos \varphi\right) + mgl \sin \varphi.$$

Problem 2, part (b) The coordinates of both points can be determined just from one angle of incline of the rigid rod connecting them φ, so the configuration manifold is $Q = S^1$. The coordinates and velocities are

$$\mathbf{x}_1 = l(\cos \varphi, 0), \quad \mathbf{x}_2 = l(0, \sin \varphi)$$
$$\mathbf{v}_1 = l\dot{\varphi}(-\sin \varphi, 0), \quad \mathbf{v}_2 = l\dot{\varphi}(0, \cos \varphi).$$

The Lagrangian is just the kinetic energy

$$L = T = \frac{1}{2}\dot{\varphi}^2 l^2 (m_1 \sin^2 \varphi + m_2 \cos^2 \varphi).$$

Momenta in terms of velocities and vice versa are computed as

$$p = \frac{\partial L}{\partial \dot{\varphi}} = \dot{\varphi}l^2(m_1 \sin^2 \varphi + m_2 \cos^2 \varphi), \quad \dot{\varphi} = \frac{p}{l^2(m_1 \sin^2 \varphi + m_2 \cos^2 \varphi)}.$$

The Hamiltonian is then given by

$$H = \frac{p^2}{2l^2(m_1 \sin^2 \varphi + m_2 \cos^2 \varphi)}.$$

We finally arrive to Hamilton's equations:

$$\dot{p} = -\frac{\partial H}{\partial \varphi} = \frac{p^2(m_1 - m_2)\sin\varphi\cos\varphi}{l^2(m_1\sin^2\varphi + m_2\cos^2\varphi)^2}$$

$$\dot{\varphi} = \frac{\partial H}{\partial p} = \frac{p}{l^2(m_1\sin^2\varphi + m_2\cos^2\varphi)^2}.$$

Note that if $m_1 = m_2$, then $\dot{p} = 0$ and $\dot{\varphi} =$const, so the rod is rotating at a constant rate.

Problem 3

1. $F = p^2 + q^2$, $G = \arctan(p/q)$:

$$\{F, G\} = 2q\frac{1/q}{1 + p^2/q^2} - 2p\frac{-p/q^2}{1 + p^2/q^2} = 2$$

2. $F = p_1^2 + q_2^2$, $G = p_2^2 + q_1^2 + \varphi(\xi)$, $\xi = q_1^2 + q_2^2$:

$$\{F, G\} = \begin{pmatrix} 0 \\ 2q_2 \end{pmatrix} \cdot \begin{pmatrix} 0 \\ 2p_2 \end{pmatrix} - \begin{pmatrix} 2p_1 \\ 0 \end{pmatrix} \cdot \begin{pmatrix} 2q_1(1 + \varphi'(\xi)) \\ 2q_2\varphi'(\xi) \end{pmatrix}$$

$$= 4p_2q_2 - 4p_1q_1(1 + \varphi'(\xi))$$

3. $F = F(\mathbf{q}, \mathbf{p})$, $G = \varphi(F)$:

$$\{F, G\} = \frac{\partial F}{\partial \mathbf{q}} \cdot \varphi'(F)\frac{\partial F}{\partial \mathbf{p}} - \frac{\partial F}{\partial \mathbf{p}} \cdot \varphi'(F)\frac{\partial F}{\partial \mathbf{q}} = 0.$$

Problem 4

1. $H = F(\varphi_1(p_1, q_1), \varphi_2(p_2, q_2) \ldots, \varphi_n(p_n, q_n))$. Since φ_i only depends on (p_i, q_i), then the only derivatives in $\{\varphi_i, H\}$ should be taken with respect to p_i and q_i. Then,

$$\{\varphi_i, H\} = \frac{\partial\varphi_i}{\partial q_i}\frac{\partial F}{\partial\varphi_i}\frac{\partial\varphi_i}{\partial p_i} - \frac{\partial\varphi_i}{\partial p_i}\frac{\partial F}{\partial\varphi_i}\frac{\partial\varphi_i}{\partial q_i} = 0.$$

2. We need to consider

$$P_i = f_i - H\varphi_i, \quad H = \frac{\sum_{i=1}^n f_i(p_i, q_i)}{\sum_{i=1}^n \varphi_i(p_i, q_i)}$$

Since $\{f_i, f_i\} = \{\varphi_i, \varphi_i\} = \{H, H\} = 0$, we proceed as follows:

$$\{f_i, H\} = -\frac{\{f_i, \varphi_i\}\sum_{i=1}^n f_i}{\left(\sum_{i=1}^n \varphi_i\right)^2} = -H\frac{\{f_i, \varphi_i\}}{\sum_{i=1}^n \varphi_i}$$

$$\{\varphi_i, H\} = \frac{\{\varphi_i, f_i\}}{\sum_{i=1}^n \varphi_i} = -\frac{\{f_i, \varphi_i\}}{\sum_{i=1}^n \varphi_i}$$

$$\{P_i, H\} = 0.$$

Thus, P_i are constants of motion of the given Hamiltonian.

A.7 Chapter 7

Problem 1

1. If

$$X = \xi^1 \frac{\partial}{\partial \xi^1} + \xi^2 \frac{\partial}{\partial \xi^2} + \xi^3 \frac{\partial}{\partial \xi^3}, \quad Y = -\xi^2 \frac{\partial}{\partial \xi^1} + \xi^1 \frac{\partial}{\partial \xi^2} + (\xi^1 + \xi^2)\frac{\partial}{\partial \xi^3}.$$

Then

$$X(Y) = \xi^1 \left(\frac{\partial}{\partial \xi_2} + \frac{\partial}{\partial \xi_3}\right) + \xi^2 \left(-\frac{\partial}{\partial \xi_1} + \frac{\partial}{\partial \xi_3}\right) + \text{2nd order}$$

$$Y(X) = -\xi^2 \frac{\partial}{\partial \xi_1} + \xi^1 \frac{\partial}{\partial \xi_2} + (\xi^1 + \xi^2)\frac{\partial}{\partial \xi^3} + \text{2nd order}$$

$$X(Y) - Y(X) = 0.$$

2. Here, we proceed simply by definition:

$$f = (\xi^1 + \xi^2)^2 + (\xi^3)^2$$
$$df = 2(\xi^1 + \xi^2)d\xi^1 + 2(\xi^1 + \xi^2)d\xi^2 + 2\xi^3 d\xi^3$$
$$i_X df = 2(\xi^1 + \xi^2)d\xi^1(X) + 2(\xi^1 + \xi^2)d\xi^2(X) + 2\xi^3 d\xi^3(X)$$
$$= 2(\xi^1 + \xi^2)\xi^1 + 2(\xi^1 + \xi^2)\xi^2 + 2\xi^3\xi^3 = 2f \Rightarrow \text{Interesting!}$$
$$i_Y df = 2(\xi^1 + \xi^2)d\xi^1(Y) + 2(\xi^1 + \xi^2)d\xi^2(Y) + 2\xi^3 d\xi^3(Y)$$
$$= -2\xi^2(\xi^1 + \xi^2) + 2(\xi^1 + \xi^2)\xi^1 + 2\xi^3(\xi^1 + \xi^2)$$
$$= 2(\xi^1 + \xi^2)(\xi^1 - \xi^2 + \xi^3).$$

3. If $\theta = \xi^1 d\xi^1 + (\xi^2 - \xi^3)d\xi^2 + (\xi^3 - \xi^2)d\xi^3$, then

$$d\theta = -d\xi^3 \wedge d\xi^2 - d\xi^2 \wedge d\xi^3 = 0,$$

so the form is closed. If the form is exact, then $\theta = dF$ for some function F. We see that we would need

$$\xi^1 = \frac{\partial f}{\partial \xi^1}, \quad (\xi^2 - \xi^3) = \frac{\partial f}{\partial \xi^2}, \quad (\xi^3 - \xi^2) = \frac{\partial f}{\partial \xi^3},$$

which is satisfied for $f = \frac{1}{2}(\xi^1)^2 + \frac{1}{2}(\xi^2 - \xi^3)^2$, so the form is also exact. In fact, the difference between the exact and closed forms plays an important role in describing the topology of the space; see [1].

4. This problem is solved directly by definition:

$$\theta = \xi^3 d\xi^1 + (\xi^2 - \xi^1)d\xi^2 + (\xi^3 - \xi^2)d\xi^3$$
$$d\theta = d\xi^3 \wedge d\xi^1 - d\xi^1 \wedge d\xi^2 - d\xi^2 \wedge d\xi^3$$
$$= -d\xi^1 \wedge d\xi^2 - d\xi^1 \wedge d\xi^3 - d\xi^2 \wedge d\xi^3$$
$$i_X d\theta = -d\xi^1(X)d\xi^2 + d\xi^1 d\xi^2(X) - d\xi^1(X)d\xi^3 + d\xi^1 d\xi^3(X)$$
$$\qquad - d\xi^2(X)d\xi^3 + d\xi^2 d\xi^3(X)$$
$$= -\xi^1 d\xi^2 + d\xi^1 \xi^2 - \xi^1 d\xi^3 + d\xi^1 \xi^3 - \xi^2 d\xi^3 + d\xi^2 \xi^3$$
$$= (\xi^2 + \xi^3)d\xi^1 + (-\xi^1 + \xi^3)d\xi^2 - (\xi^1 + \xi^2)d\xi^3$$
$$i_Y i_X d\theta = -\xi^2(\xi^2 + \xi^3) + (-\xi^1 + \xi^3)\xi^1 - (\xi^1 + \xi^2)^2$$
$$i_X \theta = \xi^3 \xi^1 + (\xi^2 - \xi^1)\xi^2 + (\xi^3 - \xi^2)\xi^3$$
$$di_X \theta = \xi^1 d\xi^3 + \xi^3 d\xi^1 + \xi^2(d\xi^2 - d\xi^1) + (\xi^2 - \xi^1)d\xi^2 + \xi^3(d\xi^3 - d\xi^2)$$
$$\qquad + (\xi^3 - \xi^2)d\xi^3$$
$$= (\xi^3 - \xi^2)d\xi^1 + (2\xi^2 - \xi^1 - \xi^3)d\xi^2 + (\xi^1 - \xi^2 + 2\xi^3)d\xi^3.$$

Problem 2

1. If $\theta = \theta_1 d\xi^1 + \theta_2 d\xi^2 + \theta_3 d\xi^3$, then

$$d\theta = \frac{\partial \theta_1}{\partial \xi^2}d\xi^2 \wedge d\xi^1 + \frac{\partial \theta_1}{\partial \xi^3}d\xi^3 \wedge d\xi^1 + \frac{\partial \theta_2}{\partial \xi^1}d\xi^1 \wedge d\xi^2 + \frac{\partial \theta_2}{\partial \xi^3}d\xi^3 \wedge d\xi^2$$
$$+ \frac{\partial \theta_3}{\partial \xi^1}d\xi^1 \wedge d\xi^3 + \frac{\partial \theta_3}{\partial \xi^2}d\xi^2 \wedge d\xi^3$$

$$= \left(\frac{\partial \theta_2}{\partial \xi^1} - \frac{\partial \theta_1}{\partial \xi^2} \right) d\xi^1 \wedge d\xi^2 + \left(\frac{\partial \theta_1}{\partial \xi^3} - \frac{\partial \theta_3}{\partial \xi^1} \right) d\xi^3 \wedge d\xi^1$$

$$+ \left(\frac{\partial \theta_3}{\partial \xi^2} - \frac{\partial \theta_2}{\partial \xi^3} \right) d\xi^2 \wedge d\xi^3.$$

Thus, the coefficients of $d\xi^2 \wedge d\xi^3, d\xi^3 \wedge d\xi^1, d\xi^1 \wedge d\xi^2$ in $d\theta$ correspond to the coefficients of curl θ.

2. If $\theta = dF$, then $\theta_\alpha = \frac{\partial F}{\partial \xi^\alpha}$. In turn, from the exterior calculus, we know that $d\theta = ddF = 0$. On the other hand, the coefficients of $d\theta$, as we have seen, will be the coefficients of curl θ, and if $\theta = \nabla F$, then we get curl grad $F = 0$.

3. If

$$\theta = v_1 d\xi^2 \wedge d\xi^3 + v_2 d\xi^3 \wedge d\xi^1 + v_3 d\xi^1 \wedge d\xi^2,$$

then

$$d\theta = \frac{\partial v_1}{\partial \xi^1} d\xi^1 \wedge d\xi^2 d\xi^3 + \frac{\partial v_2}{\partial \xi^2} d\xi^2 \wedge d\xi^3 \wedge d\xi^1 + \frac{\partial v_3}{\partial \xi^3} d\xi^3 d\xi^1 \wedge d\xi^2$$

$$= \text{div } v d\xi^1 \wedge d\xi^2 d\xi^3.$$

Problem 3 Let us introduce the following notation:

$$\theta = \theta_\alpha d\xi^\alpha, \quad X = X^\beta \frac{\partial}{\partial \xi^\beta}, \quad Y = Y^\beta \frac{\partial}{\partial \xi^\beta}.$$

Then,

$$d\theta = \frac{\partial \theta_\alpha}{\partial \xi^s} d\xi^s \wedge d\xi^\alpha$$

$$i_Y d\theta = \frac{\partial \theta_\alpha}{\partial \xi^s} \left(d\xi^s(Y) d\xi^\alpha - d\xi^\alpha(Y) d\xi^s \right) = \frac{\partial \theta_\alpha}{\partial \xi^s} \left(Y^s d\xi^\alpha - Y^\alpha d\xi^s \right)$$

$$i_X i_Y d\theta = \frac{\partial \theta_\alpha}{\partial \xi^s} \left(Y^s X^\alpha - Y^\alpha X^s \right)$$

$$i_X \theta = \theta_\alpha X^\alpha$$

$$i_Y \theta = \theta_\alpha Y^\alpha$$

$$d(i_X \theta) = \frac{\partial \theta^\alpha}{\partial \xi^s} X^\alpha d\xi^s + \theta^\alpha \frac{\partial X^\alpha}{\partial \xi^s} d\xi^s$$

$$d(i_Y \theta) = \frac{\partial \theta^\alpha}{\partial \xi^s} Y^\alpha d\xi^s + \theta_\alpha \frac{\partial Y^\alpha}{\partial \xi^s} d\xi^s.$$

From these equations, we conclude that

$$i_X d(i_Y \theta) - i_X d(i_Y \theta) = \theta_\alpha \left(\frac{\partial Y^\alpha}{\partial \xi^\beta} X^\beta - \frac{\partial X^\alpha}{\partial \xi^\beta} Y^\beta \right) + \frac{\partial \theta_\alpha}{\partial \xi^\beta} \left(Y^\alpha X^\beta - X^\alpha Y^\beta \right).$$

Thus, we see that

$$i_X d(i_Y \theta) - i_Y d(i_X \theta) = i_{[X,Y]} \theta + i_X i_Y d\theta$$

and we conclude

$$i_Y i_X d\theta = d\theta(X, Y) = i_X d(i_Y \theta) - i_Y d(i_X \theta) - i_{[X,Y]} \theta.$$

Since X and Y are arbitrary, this identity is an alternative way of computing $d\theta$ when θ is a one-form.

A.8   Chapter 8

Problem 1 Since we are dealing with one-dimensional system, $\{Q, Q\} = \{P, P\} = 0$. Thus, we only need to check the condition $\{Q, P\} = 1$. Since $Q = \frac{1}{2}(q^2 + p^2)$, the Poisson bracket of Q with an arbitrary function of $q^2 + p^2$ will vanish. Then, we take Φ to be a function of $r = \sqrt{p^2 + q^2}$ and $\xi = p/q$ (we can, of course, take $\xi = \arctan p/q$, but the formulas are easier without the arc tan). Thus,

$$\{Q, \Phi\} = \frac{\partial \Phi}{\partial \xi} \{Q, \xi\} = \frac{\partial \Phi}{\partial \xi} \left(q \frac{1}{q} + p \frac{p}{q^2} \right) = \frac{\partial \Phi}{\partial \xi} (1 + \xi^2) = 1.$$

Thus,

$$\Phi(\xi, r) = f(r) + \int \frac{d\xi}{1 + \xi^2} = f(r) + \arctan \xi,$$

where f is an arbitrary function of r. One can also write this expression as

$$\Phi(p, q) = F(q^2 + p^2) + \arctan \frac{p}{q},$$

where $F(x)$ is an arbitrary function of its argument $x = q^2 + p^2$.

Problem 2 Since every Q_i and P_i depends only on (q_i, p_i) for every i, we only need to verify the identities for every i, since $\{Q_i, Q_j\} = \{P_i, P_j\} = \{Q_i, P_j\} = 0$ when $i \neq j$. Thus, we only need to check whether $\{Q_i, P_i\} = 1$, since $\{Q_i, Q_i\} = \{P_i, P_i\} = 0$. By definition,

$$\{Q_i \, P_i\} = \{q_i - \gamma_i t \log(p_i \gamma_i t), \, p_i\} = \{q_i, p_i\} = 1 .$$

Thus, the transformation is canonical.

Problem 3

1. We notice again that $Q_i = \arctan(p_i q_i)$, $P_i = -\left(1 + (p_i q_i)^2\right) \log q_i$ only depend on pairs (q_i, p_i) and no other variables. Thus, $\{Q_i, Q_j\} = \{P_i, P_j\} = 0$ when $i \neq j$. Since $\{Q_i, Q_i\} = \{P_i, P_i\} = 0$, we only need to verify the identity for $\{Q_i, P_i\}$. Since Q_i depends on the combination of variables $q_i p_i$, and P_i have a component depending on the same combination $q_i p_i$, we compute

$$\{Q_i, P_i\} = -\left(1 + (p_i q_i)^2\right)\{\arctan(p_i q_i), \log q_i\}$$

$$= \left(1 + (p_i q_i)^2\right) \frac{q_i}{1 + (p_i q_i)^2} \frac{1}{q_i} = 1.$$

2. When we take $(\mathbf{q}, \mathbf{Q})$ representation, we need to express $\mathbf{P} = \mathbf{P}(\mathbf{q}, \mathbf{Q})$ and $\mathbf{p} = \mathbf{p}(\mathbf{q}, \mathbf{Q})$. This transformation is written as

$$P_i = -(1 + (\tan Q_i)^2) \log q_i , \quad p_i = \frac{\tan Q_i}{q_i}$$

If the generating function is $F(\mathbf{q}, \mathbf{Q})$, then

$$P_i = -(1 + (\tan Q_i)^2) \log q_i = -\frac{\log q_i}{\cos^2 Q_i} = -\frac{\partial F}{\partial Q_i} .$$

$$p_i = \frac{\tan Q_i}{q_i} = \frac{\partial F}{\partial q_i}$$

From the p_i equation, we get $F = \tan Q_i \log q_i + \varphi(Q_i)$; substitution of this function into the P_i yields $\varphi = 0$ (or constant), so $F = \tan Q_i \log q_i$.

Problem 4

1. Using the definition of the type 3 generating function, we get

$$q_i = -\frac{\partial F_3}{\partial p_i} = -C_i \frac{1}{p_i} \log(t Q_i), \quad P_i = -\frac{\partial F_3}{\partial Q_i} = -C_i \frac{1}{Q_i} \log(t p_i).$$

2. We can invert the functions from the first equation using $q_i p_i = -C_i \log(t Q_i)$, so

$$Q_i(\mathbf{q}, \mathbf{p}, t) = \frac{1}{t} e^{-\frac{p_i q_i}{C_i}}, \quad P_i(\mathbf{q}, \mathbf{p}, t) = -C_i t e^{\frac{p_i q_i}{C_i}} \log(t p_i).$$

Clearly, each Q_i and P_i is only dependent on (p_i, q_i).

A.9   Chapter 9

Problem 1

1. We compute the Hamiltonian first. Since

$$L = \frac{m}{2} \left(\dot{r}^2 + r^2 \dot{\theta}^2 + \dot{\varphi}^2 r^2 \sin^2 \theta \right) - \dot{\varphi} \lambda \cos \theta,$$

the expressions of momenta and the corresponding expressions of momenta through velocities are

$$
\begin{cases}
p_r = \dfrac{\partial L}{\partial \dot{r}} = m \dot{r} \\[2mm]
p_\theta = \dfrac{\partial L}{\partial \dot{\theta}} = m r^2 \dot{\theta} \\[2mm]
p_\varphi = m r^2 \dot{\varphi} \sin^2 \theta - \lambda \cos \theta
\end{cases}
\Rightarrow
\begin{cases}
\dot{r} = \dfrac{p_r}{m} \\[2mm]
\dot{\theta} = \dfrac{p_\theta}{m r^2} \\[2mm]
\dot{\varphi} = \dfrac{p_\varphi + \lambda \cos \theta}{m r^2 \sin^2 \theta}
\end{cases}
$$

The Hamiltonian is computed after some algebra as

$$H = p_r \dot{r} + p_\theta \dot{\theta} + p_\varphi \dot{\varphi} - L = \frac{p_r^2}{2m} + \frac{p_\theta^2}{2mr^2} + \frac{(p_\varphi + \lambda \cos \theta)^2}{2mr^2 \sin^2 \theta}. \tag{A.60}$$

2. Since H defined by Eq. (A.60) is independent of time, we look for a solution of Hamilton-Jacobi equation as $H = -Et + W(r, \theta, \varphi)$, where

$$E = \frac{1}{2m} \left(\frac{\partial W}{\partial r} \right)^2 + \frac{1}{2mr^2} \left(\frac{\partial W}{\partial \theta} \right)^2 + \frac{1}{2mr^2 \sin^2 \theta} \left(\frac{\partial W}{\partial \varphi} + \lambda \cos \theta \right)^2. \tag{A.61}$$

We look for solution in the form $W = W_r(r) + W_\theta(\theta) + W_\varphi(\varphi)$, leading to

$$E = \frac{1}{2m} \left(W_r'(r) \right)^2 + \frac{1}{2mr^2} \left(W_\theta'(\theta) \right)^2 + \frac{1}{2mr^2 \sin^2 \theta} \left(W_\varphi'(\varphi) + \lambda \cos \theta \right)^2. \tag{A.62}$$

We see that for consistency, we need

$$\frac{dW_\varphi}{d\varphi} = \text{const} = Q_\varphi$$

$$\left(\frac{dW_\theta}{d\theta}\right)^2 + \frac{1}{\sin^2\theta}\left(Q_\varphi + \lambda\cos\theta\right)^2 = \text{const} = Q_\theta$$

$$\left(\frac{dW_r}{dr}\right)^2 + \frac{1}{r^2}Q_\theta^2 = \text{const} = Q_r \qquad\qquad \text{(A.63)}$$

$$E = \frac{Q_r}{2m}.$$

We introduced the definitions of constants Q_r, Q_θ, and Q_φ, so the expressions are easier to compute; one can define these constants in any other way, and the result will be the same just with re-parameterization of these constants. We can solve Eq. (A.63) to get

$$W_\varphi = W_\varphi(\varphi, Q_\varphi) = Q_\varphi\varphi,$$

$$W_\theta = W_\varphi(\theta, Q_\theta, Q_\varphi) = \int_{\theta_0}^{\theta}\sqrt{Q_\theta - \frac{1}{\sin^2\theta}\left(Q_\varphi + \lambda\cos\theta\right)^2}d\theta, \qquad \text{(A.64)}$$

$$W_r = W_\varphi(r, Q_r, Q_\theta) = \int_{r_0}^{r}\sqrt{Q_r - \frac{1}{r^2}Q_\theta^2}\,dr.$$

The integrals above can be, in principle, computed, but that is not necessary. The lower limits in the integrals can be chosen arbitrarily, as long as there are no singularities in the integrals, so for example, $r_0 > 0$. We just assume that these functions are known as the quadratures and compute

$$F = -\frac{Q_r}{2m}t + W_\varphi(\varphi, Q_\varphi) + W_\varphi(\theta, Q_\theta, Q_\varphi) + W_\varphi(r, Q_r, Q_\theta). \qquad \text{(A.65)}$$

Since F defines the type 1 canonical transformation, the solutions for equations of motion obtained from Eq. (A.65) are

$$p_r = \frac{\partial F}{\partial r} = \frac{dW_r}{dr} = \sqrt{Q_r - \frac{1}{r^2}Q_\theta^2},$$

$$p_\theta = \frac{\partial F}{\partial\theta} = \sqrt{Q_\theta - \frac{1}{\sin^2\theta}\left(Q_\varphi + \lambda\cos\theta\right)^2},$$

$$p_\varphi = \frac{\partial F}{\partial\varphi} = Q_\varphi, \qquad\qquad\qquad\qquad\qquad\qquad\qquad\qquad \text{(A.66)}$$

$$P_r = -\frac{\partial F}{\partial Q_r} = t\frac{\partial E}{\partial Q_r} - \frac{\partial W_r}{\partial Q_r} = \frac{t}{2m} - \frac{\partial W_r}{\partial Q_r}(r, Q_r, Q_\theta) = \text{const},$$

$$P_\varphi = -\frac{\partial F}{\partial Q_\theta} = t\frac{\partial E}{\partial Q_\theta} - \frac{\partial W_r}{\partial Q_\theta} - \frac{\partial W_\theta}{\partial Q_\theta} = -\frac{\partial W_r}{\partial Q_\theta} - \frac{\partial W}{\partial Q_\theta} = \text{const},$$

$$P_\theta = -\frac{\partial F}{\partial Q_\varphi} = t\frac{\partial E}{\partial Q_\varphi} - \frac{\partial W_\theta}{\partial Q_\varphi} - \frac{\partial W_\varphi}{\partial Q_\varphi} = -\frac{\partial W_\theta(\theta)}{\partial Q_\varphi} - \varphi = \text{const}.$$

The first three equations of Eq. (A.66) are explicit expressions connecting the momenta and coordinates; in particular, the third equation states that p_φ =const. The last three equations determine, implicitly, (r, θ, φ) as functions of time. In particular, the very last equation expresses φ as a function of θ. If one could find explicit expressions for the integrals in Eq. (A.64) and compute the derivatives, then one would also be able to compute the momenta from the first three equations. Implicit expressions like Eq. (A.66) are OK, though; it is often all one can hope for.

Problem 2

1. We look for a solution to the Hamilton-Jacobi equation in the following form:

$$F = W_0(t) + \sum_\alpha W_\alpha(q^\alpha).$$

Then, the equations for each component are given as

$$\left(\frac{dW_\alpha}{dq^\alpha}\right)^2 + \gamma_\alpha(q^\alpha)^2 = \gamma_\alpha Q^\alpha, \quad F_0'(t) = \sqrt{\sum_\alpha \gamma_\alpha Q^\alpha + \varphi(t)}.$$

Note that we have defined the constant $\gamma_\alpha Q^\alpha$ rather than Q^α, so the final expressions are a bit easier.

The solutions for $W_\alpha(q^\alpha)$ and $F_0(t)$ are

$$W_\alpha = \sqrt{\gamma_\alpha} \int_0^{q^\alpha} \sqrt{Q^\alpha - (q^\alpha)^2} dq^\alpha$$

$$= \frac{\sqrt{\gamma_\alpha}}{2}\left(q^\alpha\sqrt{Q^\alpha - (q^\alpha)^2} + Q^\alpha \arctan\frac{q^\alpha}{\sqrt{Q^\alpha - (q^\alpha)^2}}\right),$$

$$F_0(t) = F_0(t, \mathbf{Q}) = \int_0^t \sqrt{\sum_\alpha \gamma_\alpha Q^\alpha + \varphi(t)}\, dt.$$

The explicit solution for W_α is actually not that important; the fact that the solution integrates out in elementary functions is, of course, nice. For more complex Hamiltonians, you would have to deal with expressions in quadratures.

The solutions for p_α and q_α are, after taking the derivatives,

$$p_\alpha = \frac{\partial F}{\partial q^\alpha} = \frac{dW_\alpha}{dq^\alpha} = \sqrt{\gamma_\alpha}\sqrt{Q^\alpha - (q^\alpha)^2},$$

$$P_\alpha = -\frac{\partial F}{\partial Q^\alpha} = -\int_0^t \frac{\gamma_\alpha dt}{\sqrt{\gamma_\alpha Q^\alpha + \varphi(t)}} - \frac{\sqrt{\gamma}}{2} \arctan \frac{q^\alpha}{\sqrt{Q^\alpha - (q^\alpha)^2}} = \text{const.}$$

$$(\text{A.67})$$

While it is not always possible, in this case, we can proceed a bit further. We can actually find q^α explicitly as a function of time if we assume $\frac{\partial F_0}{\partial Q_\alpha}$ to be a known function. If we define

$$\Psi_\alpha(t) = P_\alpha + \int_0^t \frac{\gamma_\alpha dt}{\sqrt{\gamma_\alpha Q^\alpha + \varphi(t)}},$$

then

$$\frac{q^\alpha}{\sqrt{Q^\alpha - (q^\alpha)^2}} = \tan \Psi_\alpha \quad \Rightarrow \quad q^\alpha = \sqrt{Q^\alpha} \sin \Psi_\alpha(t), \; p_\alpha = \sqrt{\gamma_\alpha Q^\alpha} \cos \Psi_\alpha(t),$$

where the momenta are computed from Eq. (A.67). But, in general, it is perfectly OK to leave the solution in the form Eq. (A.67).

Problem 3 Let us write q and Q with indices down, so we don't confuse indices with powers. Since the Hamiltonian is independent of time, we look for solutions in the following form:

$$F = -Et + \sum_{i=1}^3 W_i(q_i),$$

with the condition on $W_i(q_i)$:

$$E = \frac{1}{q_1^2}\left(\frac{dW_1}{dq_1}\right)^2 + \frac{1}{q_1^2}\left(\frac{1}{q_2^2}\left(\frac{dW_2}{dq_2}\right)^2 + \frac{1}{q_2^2 q_3^2}\left(\frac{dW_3}{dq_3}\right)^2\right),$$

and we get the solution

$$W_3(q_3) = \frac{1}{2}Q_3 q_3^2,$$

$$W_2(q_2) = \int_0^{q_2} \sqrt{Q_2 q_2^2 - Q_3} dq_2 = \Phi(q_2; Q_2, Q_3),$$

$$W_1(q_1) = \int_0^{q^1} \sqrt{Q_1 q_1^2 - Q_2} dq_1 = \Phi(q_1; Q_1, Q_2), \quad \text{where}$$

$$\Phi(x, A, B) := \int_0^x \sqrt{Ax^2 - B} dx$$

$$= \frac{1}{2}\left(x\sqrt{Ax^2 - x} - \frac{B}{\sqrt{A}} \log\left[Ax + \sqrt{A}\sqrt{Ax^2 - B}\right]\right),$$

$$E = Q_1.$$

It is OK to use symbolic computations in computing F. The derivatives of F with respect to A and B are a bit convoluted, so we just keep them implicitly here. Of course, you can use symbolic calculations to compute them in explicit form. The semicolon in $F(x; A, B)$ indicates that A and B are parameters that are assumed to be constant. Then, the solution is

$$F = -Q_1 t + \Phi(q_1; Q_1, Q_2) + \Phi(q_2; Q_2, Q_3) + \frac{1}{2}Q_3 q_3^2.$$

$$p_1 = \frac{\partial F}{\partial q_1} = \sqrt{Q_1 q_1^2 - Q_2},$$

$$p_2 = \frac{\partial F}{\partial q_2} = \sqrt{Q_1 q_2^2 - Q_3},$$

$$p_3 = \frac{\partial F}{\partial q^3} = Q_3 q_3,$$

$$P_1 = -\frac{\partial F}{\partial Q_1} = t - \frac{\partial \Phi}{\partial A}(q_1; Q_1, Q_2) = \text{const},$$

$$P_2 = -\frac{\partial F}{\partial Q_2} = -\frac{\partial \Phi}{\partial B}(q_1; Q_1, Q_2) - \frac{\partial \Phi}{\partial A}(q_2; Q_2, Q_3) = \text{const},$$

$$P_3 = -\frac{\partial \Phi}{\partial B}(q_2; Q_2, Q_3) + \frac{1}{2}q_3^2 = \text{const}.$$

$$(A.68)$$

The first three equations of Eq. (A.68) determine connections between p_α and q_α. The fourth equation of Eq. (A.68) defines q_1 as a function of time, the fifth equation determines q_2 as a function of q_1, and the last equation connects q_3 with q_2.

Problem 4 We remember that for the case where the dimension of both q and p is one, the action variable has a very simple geometric meaning; it is simply the value of the area bounded by the solution curve on the plane. The curve specified by $H = $ const is an ellipse, with the semiaxes in the p-direction $a = \sqrt{2mH}$ and in the q-direction $b = \sqrt{2H/c}$. The area of that ellipse is

$$J = \pi ab = 2\pi H \sqrt{\frac{m}{c}}, \quad \Leftrightarrow \quad H = \frac{1}{2\pi}\sqrt{\frac{c}{m}} J. \tag{A.69}$$

We know that the mapping from (q, p) to (J, φ) must be a canonical transformation. Since we know J, let us define this transformation using the type 1 generating function $F(q, Q)$, where $Q = J$. Then, we know that

$$p = \sqrt{m(2H - cq^2)} = \sqrt{\frac{\alpha}{\pi}J - \alpha^2 q^2} = \frac{\partial F}{\partial q}, \quad \text{where} \quad \alpha := \sqrt{mc}.$$

Then, we compute the generating function as

$$F = \int^q p(q, J)dq = \int^q \sqrt{\frac{\alpha}{\pi}J - \alpha^2 q^2}dq.$$

We are not going to perform that integration, although F can be computed in terms of elementary functions. Instead, observe that the derivative is given by

$$\frac{\partial F}{\partial J} = \frac{\alpha}{2\pi}\int^q \frac{dq}{\sqrt{\frac{\alpha}{\pi}J - \alpha^2 q^2}} = \frac{1}{2\pi}\arctan\frac{\alpha q}{\sqrt{\frac{\alpha}{\pi}J - \alpha^2 q^2}}.$$

Thus, the angle variable is

$$\varphi = P = -\frac{\partial F}{\partial J} = \frac{1}{2\pi}\arctan\frac{\alpha q}{\sqrt{\frac{\alpha}{\pi}J - \alpha^2 q^2}} = -\frac{1}{2\pi}\arctan\frac{\alpha q}{p(q, J)},$$

so $\varphi(J)$ is connected to the angle of the point at the (q, p) plane.

Since the generating function F does not depend on time $\partial_t F = 0$, the new Hamiltonian K is simply H expressed in the new variables according to Eq. (A.69). The equations of motion in the action-angle variables are

$$\dot{\varphi} = -\frac{\partial K}{\partial J} = \frac{1}{2\pi}\sqrt{\frac{c}{m}} = \text{const}, \quad \dot{J} = \frac{\partial K}{\partial \varphi} = 0. \tag{A.70}$$

Problem 5

(a) The shape of the curve in the (p, q) plane for the fixed slow speed is a rectangle with the area given as $J = 2pL$, which is adiabatic invariant.

(b) The Hamiltonian is given by

$$H = \frac{p^2}{2m} - l \cos q,$$

so for a fixed energy, $p = \pm\sqrt{2m(H + l \cos q)}$, with $p = 0$ at $q = q_0$ (max amplitude). The area bounded by the curve is

$$J = 2 \int_{-q_0}^{q_0} \sqrt{2m(H + l \cos q)}\,dq = 4 \int_0^{q_0} \sqrt{2m(H + l \cos q)}\,dq.$$

This integral can be computed in terms of elliptic functions, but we won't do it here.

A.10   Chapter 10

Problem 1

(1a) The rotation matrices about the axis $\mathbf{E_1} = \mathbf{e}_1$ are

$$R_1(\varphi_1) = \begin{pmatrix} 1 & 0 & 0 \\ 0 & \cos \varphi_1 & -\sin \varphi_1 \\ 0 & \sin \varphi_1 & \cos \varphi_1 \end{pmatrix}, \quad R_2(\varphi_2) = \begin{pmatrix} \cos \varphi_2 & 0 & -\sin \varphi_2 \\ 0 & 1 & 0 \\ \sin \varphi_2 & 0 & \cos \varphi_2 \end{pmatrix}, \tag{A.71}$$

and the product is

$$R_2(\varphi_2)R_1(\varphi_1) = \begin{pmatrix} \cos \varphi_2 & -\sin \varphi_1 \sin \varphi_2 & -\cos \varphi_1 \sin \varphi_2 \\ 0 & \cos \varphi_1 & -\sin \varphi_1 \\ \sin \varphi_2 & \sin \varphi_1 \cos \varphi_2 & \cos \varphi_1 \cos \varphi_2 \end{pmatrix}.$$

(1b) Using Eq. (A.71), we compute the product

$$R_1(\theta)R_1(\varphi) = \begin{pmatrix} 1 & 0 & 0 \\ 0 & \cos \varphi \cos \theta - \sin \varphi \sin \theta & -\sin \varphi \cos \theta - \cos \varphi \sin \theta \\ 0 & \sin \varphi \cos \theta + \cos \varphi \sin \theta & \cos \varphi \cos \theta - \sin \varphi \sin \theta \end{pmatrix},$$

which simplifies as

$$R_1(\theta)R_1(\varphi) = \begin{pmatrix} 1 & 0 & 0 \\ 0 & \cos(\varphi + \theta) & -\sin(\varphi + \theta) \\ 0 & \sin(\varphi + \theta) & \cos(\varphi + \theta) \end{pmatrix} = R_1(\varphi)R_1(\theta).$$

(1c) Take the axis of rotation to be, for example, along the axis $\mathbf{e}_1$, and consider $\mathbb{R}_1(\varphi(t))$ where $\mathbb{R}_1(\varphi)$ is given by Eq. (A.71). Then,

$$\mathbb{R}_1(\varphi(t)) = \begin{pmatrix} 1 & 0 & 0 \\ 0 & \cos\varphi(t) & -\sin\varphi(t) \\ 0 & \sin\varphi(t) & \cos\varphi(t) \end{pmatrix},$$

$$\dot{\mathbb{R}}_1(\varphi(t)) = \begin{pmatrix} 0 & 0 & 0 \\ 0 & -\sin\varphi(t) & -\cos\varphi(t) \\ 0 & \cos\varphi(t) & -\sin\varphi(t) \end{pmatrix} \dot\varphi$$

Then, a direct calculation gives

$$\widehat{\Omega} = \mathbb{R}_1^T \dot{\mathbb{R}}_1 = \begin{pmatrix} 0 & 0 & 0 \\ 0 & 0 & -1 \\ 0 & 1 & 0 \end{pmatrix}\dot\varphi = \dot{\mathbb{R}}_1 \mathbb{R}_1^T = \widehat\omega,$$

which also means $\Omega = \omega$, and the spatial and body angular velocities are the same.

Problem 2 Consider the rigid body and write $\Omega = (0, \Omega, 0) + \epsilon(\alpha_1, \alpha_2, \alpha_3)$. Euler's equations are

$$\begin{cases} \epsilon I_1 \dot\alpha_1 = \epsilon(I_2 - I_3)\Omega\alpha_3 + O(\epsilon^2), \\ \epsilon I_2 \dot\alpha_2 = +O(\epsilon^2), & \text{(A.72)} \\ \epsilon I_3 \dot\alpha_3 = \epsilon(I_1 - I_2)\Omega\alpha_1 + O(\epsilon^2). \end{cases}$$

Differentiating the first equation and using the third equation, we obtain

$$I_1 I_3 \ddot\alpha_1 = (I_2 - I_3)(I_1 - I_2)\Omega^2\alpha_1 .$$

Since $I_1 < I_2$ and $I_2 < I_3$,

$$\ddot\alpha_1 = \lambda^2\alpha_1 , \quad \lambda_1 = \sqrt{\frac{(I_3 - I_2)(I_2 - I_1)}{I_1 I_3}}\,\Omega$$

and $\alpha_1 \sim e^{\lambda t}$, so the critical point is unstable.

Problem 3 In this problem, we need to compute

$$\dot{\Omega} = \mathbb{I}^{-1}(-\Omega \times \mathbb{I}\Omega + mg\Gamma \times A) = \Phi(\Omega, \Gamma).$$

(a) $F = \boldsymbol{\Omega} \cdot \boldsymbol{\Gamma}$:

$$\frac{dF}{dt} = \dot{\boldsymbol{\Omega}} \cdot \boldsymbol{\Gamma} + \boldsymbol{\Omega} \cdot \dot{\boldsymbol{\Gamma}} = \mathbb{I}^{-1}(-\boldsymbol{\Omega} \times \mathbb{I}\boldsymbol{\Omega} + mg\boldsymbol{\Gamma} \times A) \cdot \boldsymbol{\Gamma} + \boldsymbol{\Omega} \cdot (-\boldsymbol{\Omega} \times \boldsymbol{\Gamma})$$

$$= \mathbb{I}^{-1}(-\boldsymbol{\Omega} \times \mathbb{I}\boldsymbol{\Omega} + mg\boldsymbol{\Gamma} \times A) \cdot \boldsymbol{\Gamma}.$$

(b) $F = \boldsymbol{\Omega} \cdot (\boldsymbol{\Gamma} \times \mathbf{b})$:

$$\frac{dF}{dt} = \dot{\boldsymbol{\Omega}} \cdot (\boldsymbol{\Gamma} \times \mathbf{b}) + \boldsymbol{\Omega} \cdot (\dot{\boldsymbol{\Gamma}} \times \mathbf{b})$$

$$= \mathbb{I}^{-1}(-\boldsymbol{\Omega} \times \mathbb{I}\boldsymbol{\Omega} + mg\boldsymbol{\Gamma} \times A) \cdot (\boldsymbol{\Gamma} \times \mathbf{b}) + \boldsymbol{\Omega} \cdot (-(\boldsymbol{\Omega} \times \boldsymbol{\Gamma}) \times \mathbf{b})$$

$$= \mathbb{I}^{-1}(-\boldsymbol{\Omega} \times \mathbb{I}\boldsymbol{\Omega} + mg\boldsymbol{\Gamma} \times A) \cdot (\boldsymbol{\Gamma} \times \mathbf{b}) + (\boldsymbol{\Omega} \times \mathbf{b}) \cdot (\boldsymbol{\Omega} \times \boldsymbol{\Gamma}).$$

(c) $F = (\boldsymbol{\Omega} \times \mathbf{b}) \cdot (\boldsymbol{\Gamma} \times \mathbf{c})$:

$$\frac{dF}{dt} = (\dot{\boldsymbol{\Omega}} \times \mathbf{b}) \cdot (\boldsymbol{\Gamma} \times \mathbf{c}) + (\boldsymbol{\Omega} \times \mathbf{b}) \cdot (\dot{\boldsymbol{\Gamma}} \times \mathbf{c})$$

$$= \left[\mathbb{I}^{-1}(-\boldsymbol{\Omega} \times \mathbb{I}\boldsymbol{\Omega} + mg\boldsymbol{\Gamma} \times A) \times \mathbf{b}\right] \cdot (\boldsymbol{\Gamma} \times \mathbf{c}) + (\boldsymbol{\Omega} \times \mathbf{b}) \cdot (-(\boldsymbol{\Omega} \times \boldsymbol{\Gamma}) \times \mathbf{c})$$

$$= \mathbb{I}^{-1}(-\boldsymbol{\Omega} \times \mathbb{I}\boldsymbol{\Omega} + mg\boldsymbol{\Gamma} \times A) \cdot (\mathbf{b} \times (\boldsymbol{\Gamma} \times \mathbf{c})) + (\boldsymbol{\Omega} \times \mathbf{b}) \cdot (\mathbf{c} \times (\boldsymbol{\Omega} \times \boldsymbol{\Gamma}))$$

$$= \mathbb{I}^{-1}(-\boldsymbol{\Omega} \times \mathbb{I}\boldsymbol{\Omega} + mg\boldsymbol{\Gamma} \times A) \cdot (\boldsymbol{\Gamma}(\mathbf{b} \cdot \mathbf{c}) - \mathbf{c}(\boldsymbol{\Gamma} \cdot \mathbf{b}) - (\mathbf{c} \cdot \boldsymbol{\Omega})\boldsymbol{\Gamma} \cdot (\boldsymbol{\Omega} \times \mathbf{b}),$$

where in the last equation, we have used $\boldsymbol{\Gamma} \cdot (\boldsymbol{\Omega} \times \mathbf{b}) = \boldsymbol{\Omega}(\boldsymbol{\Gamma} \cdot \mathbf{b}) - \boldsymbol{\Gamma}(\mathbf{c} \cdot \boldsymbol{\Omega})$ (the BAC-CAB identity for double vector products).

A.11   Chapter 11

Problem 1

1. The constraint on the variations is $\delta x - y\delta z = 0$. Lagrange-d'Alembert's principle states that

$$\delta \int L + \lambda(\delta x - y\delta z)dt = 0,$$

leading to the equations

$$\begin{cases} \delta x : & \ddot{x} = \lambda, \\ \delta y : & \ddot{y} = 0, \\ \delta z : & \ddot{z} = -\lambda y. \end{cases} \tag{A.73}$$

2. If we naively substitute $\dot{x} = y\dot{z}$, we get the Lagrangian

$$\tilde{L} = \frac{1}{2}\left((y^2 + 1)\dot{z}^2 + \dot{y}^2\right)$$

with the Euler-Lagrange equations

$$\ddot{y} - \dot{z}^2 y = 0, \qquad \frac{d}{dt}\dot{z}(y^2 + 1) = 0, \tag{A.74}$$

and no equation for x; the evolution of x is computed from the constraint $\dot{x} = y\dot{z}$. Clearly, Eq. (A.74)—also called *Vakonomic equations*—are very different from Eq. (A.73). In particular, $\ddot{y}$ is conserved in Eq. (A.73) and is not conserved in Eq. (A.74).
3. In Eq. (A.73), the momentum in the x direction is $p_x = \dot{x}$, and $\dot{p}_x = \lambda \neq 0$. Similarly, $p_z = \dot{z}$ and $\dot{p}_z = -\lambda y \neq 0$. In contrast, the momentum in the y-direction $p_y = \dot{y}$ is conserved.
4. The energy is conserved:

$$\dot{E} = \dot{x}\ddot{x} + \dot{y}\ddot{y} + \dot{z}\ddot{z} = \lambda\dot{x} - \lambda y\dot{z} = \lambda(\dot{x} - y\dot{z}) = 0,$$

with the terms in the last equation cancelling due to the constraint.
5. From the y-equation of Eq. (A.73), we obtain $y(t) = y_0 + Vt$. Multiplying the first equation of that system by $y(t)$ and adding to the last equation, we get

$$\ddot{z} = -\ddot{x}y.$$

Differentiating the constraint $\dot{x} = y\dot{z}$, we substitute $\ddot{x} = \dot{y}\dot{z} + y\ddot{z}$, and substituting into above equation, we obtain

$$\ddot{z} = -y\dot{y}\dot{z} - y^2\ddot{z}, \quad \Rightarrow \quad \ddot{z}(1 + y^2) = -\dot{z}y\dot{y}.$$

Thus,

$$\frac{\ddot{z}}{\dot{z}} = -\frac{y\dot{y}}{1 + y^2}, \quad \Rightarrow \quad \dot{z}\sqrt{1 + y^2} = P = \text{const}.$$

The last integral allows us to compute $z(t)$ using the solution $y(t) = y_0 + Vt$ obtained above:

$$z(t) = \int \frac{P}{\sqrt{1 + (y_0 + Vt)^2}} dt = \frac{P}{V} \log \left(\sqrt{1 + (y_0 + Vt)^2} + y_0 + Vt \right) + z_0$$

and similarly for $x(t)$.

Problem 2

1. Since $\boldsymbol{\Omega} \cdot \mathbf{a} = 0$, and $\mathbf{a}$ is fixed, then $\dot{\boldsymbol{\Omega}} \cdot \mathbf{a} = 0$. Using the equations of motion, we obtain

$$\dot{\boldsymbol{\Omega}} = -\mathbb{I}^{-1} (\boldsymbol{\Omega} \times \mathbb{I}\boldsymbol{\Omega}) + \lambda \mathbb{I}^{-1} \mathbf{a},$$

and λ is obtained as

$$\lambda = \frac{(\boldsymbol{\Omega} \times \mathbb{I}\boldsymbol{\Omega}) \cdot \mathbb{I}^{-1} \mathbf{a}}{\mathbf{a} \cdot \mathbb{I}^{-1} \mathbf{a}},$$

since $\mathbb{I}$ and its inverse are symmetric.

2. Since $\mathbb{I}$ is symmetric, the energy evolves according to

$$\dot{E} = \boldsymbol{\Omega} \cdot \mathbb{I}\dot{\boldsymbol{\Omega}} = -\boldsymbol{\Omega} \cdot (-\boldsymbol{\Omega} \times \mathbb{I}\boldsymbol{\Omega}) + \lambda \boldsymbol{\Omega} \cdot \mathbf{a} = 0.$$

The first term vanishes due to the properties of the vector product and the second vanishes due to the constraint.

3. In coordinates, the equations of motion are

$$I_1 \dot{\Omega}_1 = (I_2 - I_3)\Omega_2\Omega_3 + \lambda,$$
$$I_2 \dot{\Omega}_2 = (I_3 - I_1)\Omega_1\Omega_3,$$
$$I_3 \dot{\Omega}_3 = (I_1 - I_2)\Omega_1\Omega_2,$$
$$\Omega_1 = 0 \quad \text{(constraint)}.$$

From the constraint condition and the last two equations, we get $\dot{\Omega}_2 = \dot{\Omega}_3 = 0$, so the angular velocities are constant: $\Omega_{2,3} = \Omega_{2,3}(0) = \text{const}$.

From the first equation and the constraint, we get $\lambda = (I_3 - I_2)\Omega_2(0)\Omega_3(0) = \text{const}$.

This solution seems too trivial; however, Suslov's problem is not so trivial if $\mathbf{a}$ is not aligned with any of the inertia axes.

Bibliography

1. Arnol'd VI. Mathematical methods of classical mechanics. vol. 60. Berlin: Springer Science & Business Media; 2013.
2. Holm DD. Geometric mechanics: Dynamics and symmetry. vol. 1. London: Imperial College Press; 2008.
3. Holm DD. Geometric mechanics: part II: rotating, translating and rolling. Singapore: World Scientific Publishing Company; 2008.
4. Marsden JE, Ratiu TS. Introduction to mechanics and symmetry: a basic exposition of classical mechanical systems. vol. 17. Berlin: Springer Science & Business Media; 2013.
5. José J, Saletan E. Classical dynamics: a contemporary approach. Maryland: American Association of Physics Teachers; 2000.
6. Landau LD, Lifshitz EM. Theoretical physics. Mechanics 1988;1:121.
7. Sommerfeld A. Lectures on theoretical physics. vol. I. New York: Academic; 1964.
8. Pyatnitskii YS, Trukhan NM, Khanukayev I, Yu, Yakovenko GN. Collection of problems in analytical mechanics (in Russian). Moscow: Fizmatlit; 2002.
9. Turner MJL. Rocket and spacecraft propulsion: principles, practice and new developments. Berlin: Springer Science & Business Media; 2008.
10. Sutton GP, Biblarz O. Rocket propulsion elements. Hoboken: John Wiley & Sons; 2011.
11. Heister SD, Anderson WE, Pourpoint TL, Joseph Cassady R. Rocket propulsion. vol. 47. Cambridge: Cambridge University Press; 2019.
12. Newton I. Philosophiae naturalis principia mathematica (mathematical principles of natural philosophy). London (1687) 1687;1987.
13. Chandrasekhar S. Newton's Principia for the common reader. Oxford: Oxford University Press; 2003.
14. Dullin HR, Waalkens H. Defect in the joint spectrum of hydrogen due to monodromy. Phys Rev Lett 2018;120(2):020507.
15. Adams JC. Explanation of the observed irregularities in the motion of uranus, on the hypothesis of disturbance by a more distant planet; with a determination of the mass, orbit, and position of the disturbing body. Mon Not R Astron Soc 1846;7:149–52.
16. Airy GB. Account of some circumstances historically connected with the discovery of the planet exterior to uranus. Mon Not R Astron Soc 1846;7:121–44.
17. Challis J. Account of observations at the cambridge observatory for detecting the planet exterior to uranus. Mon Not R Astron Soc 1846;7:145–9.
18. Galle JG. Account of the discovery of Le Verrier's planet Neptune, at Berlin, Sept. 23, 1846. Mon Not R Astron Soc 1846;7:153.

© The Author(s), under exclusive license to Springer Nature Switzerland AG 2025
V. Putkaradze, *A Concise Introduction to Classical Mechanics*,
Surveys and Tutorials in the Applied Mathematical Sciences 16,
https://doi.org/10.1007/978-3-031-84977-0

19. Lissauer JJ, Marcy GW, Bryson ST, Rowe JF, Jontof-Hutter D, Agol E, Borucki WJ, Carter JA, Ford EB, Gilliland RL, et al. Validation of Kepler's multiple planet candidates. II. Refined statistical framework and descriptions of systems of special interest. Astrophys J 2014;784(1):44.

20. Valtonen MJ, Karttunen H. The three-body problem. Cambridge: Cambridge University Press; 2006.

21. Musielak ZE, Quarles B. The three-body problem. Rep Prog Phys 2014;77(6):065901.

22. Montgomery R. N-body choreographies. Scholarpedia 2010;5(11):10666.

23. Shinbrot T, Grebogi C, Wisdom J, Yorke JA. Chaos in a double pendulum. Am J Phys 1992;60(6):491–9.

24. Arnol'd VI, Kozlov VV, Neishtadt AI, Iacob I. Mathematical aspects of classical and celestial mechanics. vol. 3. Berlin: Springer; 2006.

25. Minguzzi E. Rayleigh's dissipation function at work. Euro J Phys 2015;36(3):035014.

26. de León M, Lainz M, López-Gordón A. The geometry of Rayleigh dissipation. Preprint. arXiv:2101.09036; 2021.

27. Noether E, Tavel MA. Invariant variation problems. Preprint. physics/0503066; 2005.

28. Olver PJ. Applications of Lie groups to differential equations. vol. 107. Berlin: Springer Science & Business Media; 1993.

29. Gorni G, Zampieri G. Revisiting noether's theorem on constants of motion. J Nonlin Math Phys 2014;21(1):43–73.

30. Bravetti A, Garcia-Chung A. A geometric approach to the generalized noether theorem. J Phys A Math Theor 2021;54(9):095205.

31. Mariano PM. A certain counterpart in dissipative setting of the noether theorem with no dissipation pseudo-potentials. Phil Trans R Soc A 2023;381(2263):20220375.

32. Baez JC, Fong B. A noether theorem for markov processes. J Math Phys 2013;54(1):013301.

33. Gough JE, Ratiu TS, Smolyanov OG. Noether's theorem for dissipative quantum dynamical semi-groups. J Math Phys 2015;56(2):022108.

34. Hirsch MW, Smale S, Devaney RL. Differential equations, dynamical systems, and an introduction to chaos. Cambridge: Academic Press; 2013.

35. MacKay RS. Stability of equilibria of hamiltonian systems. In: Hamiltonian dynamical systems. Boca Raton: CRC Press; 2020. p. 137–53.

36. Moser JK. Lectures on hamiltonian systems. In: Hamiltonian dynamical systems Boca Raton: CRC Press; 2020. p. 77–136.

37. Bloch AM. Stability analysis of a rotating flexible system. Acta Appl Math 1989;15(3):211–34.

38. Bloch AM, Marsden JE. Stabilization of rigid body dynamics by the energy-casimir method. Syst Control Lett 1990;14(4):341–6.

39. Ortega J-P, Planas-Bielsa V, Ratiu TS. Asymptotic and lyapunov stability of constrained and poisson equilibria. J Differ Equ 2005;214(1):92–127.

40. Birtea P, Caşu I, Ratiu TS, Turhan M. Stability of equilibria for the free rigid body. J Nonlin Sci 2012;22(2):187–212.

41. Holm DD, Kupershmidt BA. Lyapunov stability conditions for relativistic multifluid plasma. Phys D: Nonlin Phenom 1986;18(1–3):405–9.

42. Holm DD. Hamiltonian dynamics and stability analysis of neutral electromagnetic fluids with induction. Phys D: Nonlin Phenom 1987;25(1–3):261–87.

43. Holm DD. Hamiltonian techniques for relativistic fluid dynamics and stability theory. In: Relativistic fluid dynamics: lectures given at the 1st 1987 session of the Centro Internazionale Matematico Estivo (CIME) held at Noto, Italy, May 25–June 3, 1987. Berlin: Springer; 2006. p. 65–151.

44. Tronci C, Tassi E, Morrison PJ. Energy-casimir stability of hybrid vlasov-mhd models. J Phys A: Math Theor 2015;48(18):185501.

45. Bloch AM, Krishnaprasad PS, Marsden JE, Ratiu TS. Dissipation induced instabilities. In: Annales de l'Institut Henri Poincaré C, Analyse non linéaire. vol. 11. Amsterdam: Elsevier; 1994. p. 37–90.

46. Krechetnikov R, Marsden JE. Dissipation-induced instabilities in finite dimensions. Rev Mod Phys 2007;79(2):519–53.
47. Krechetnikov R, Marsden JE. Dissipation-induced instability phenomena in infinite-dimensional systems. Arch Ration Mech Anal 2009;194:611–68.
48. Holm D, Schmah T, Stoica C. Geometric mechanics and symmetry: from finite to infinite dimensions. vol. 12. Oxford: Oxford University Press; 2009.
49. Vinogradov AM, Kupershmidt BA. The structures of hamiltonian mechanics. Russ Math Surv 1977;32(4):177.
50. Abraham R, Marsden JE. Foundations of mechanics. Number 364 in AMS Chelsea Publishing. Providence: American Mathematical Society; 2008.
51. Marsden JE. Lectures on mechanics. vol. 174. Cambridge: Cambridge University Press; 1992.
52. Makeev NN. Theoretical mechanics (in Russian). Regular and Chaotic Dynamics; 1999.
53. Weinstein A. Symplectic geometry. Bull Am Math Soc 1981;5(1):1–13.
54. Arnold VI, Arnold VI. Symplectic geometry. Berlin: Springer; 1990.
55. Libermann P, Marle C-M. Symplectic geometry and analytical mechanics. vol. 35. Berlin: Springer Science & Business Media; 2012.
56. Hairer E, Wanner G, Lubich C, Hairer E, Wanner G, Lubich C. Symplectic integration of hamiltonian systems. Geometric numerical integration: structure-preserving algorithms for ordinary differential equations; 2006. p. 179–236.
57. Arnol'd VI. Small denominators and problems of stability of motion in classical and celestial mechanics. Russ Math Surv 1963;18(6):85.
58. Siegel CL, Moser JK. Lectures on celestial mechanics. Berlin: Springer Science & Business Media; 2012.
59. Vinti JP. Orbital and celestial mechanics. vol. 177. AIAA; 1998.
60. Greydanus S, Dzamba M, Yosinski J. Hamiltonian neural networks. In: Wallach H, Larochelle H, Beygelzimer A, d'Alché-Buc F, Fox E, Garnett R, editors. Advances in Neural Information Processing Systems (NeurIPS 2019), vol. 32. Curran Associates, Inc. https://proceedings.neurips.cc/paper_files/paper/2019/file/26cd8ecadce0d4efd6cc8a8725cbd1f8-Paper.pdf.
61. Jin P, Zhang Z, Zhu A, Tang Y, Karniadakis GE. Sympnets: intrinsic structure-preserving symplectic networks for identifying hamiltonian systems. Neural Netw 2020;132:166–79.
62. Li S-H, Dong C-X, Zhang L, Wang L. Neural canonical transformation with symplectic flows. Phys Rev X 2020;10(2):021020.
63. Chen R, Tao M. Data-driven prediction of general hamiltonian dynamics via learning exactly-symplectic maps. In: International conference on machine learning. PMLR; 2021. p. 1717–27.
64. Jin P, Zhang Z, Kevrekidis IG, Karniadakis GE. Learning poisson systems and trajectories of autonomous systems via poisson neural networks. IEEE Tran Neural Netw Learn Syst 2022;34(11):8271–83.
65. Eldred C, Gay-Balmaz F, Huraka S, Putkaradze V. Lie–Poisson neural networks (LPNets): data-based computing of hamiltonian systems with symmetries. Neural Netw 2024;173:106162.
66. Eldred C, Gay-Balmaz F, Putkaradze V. CLPNets: coupled lie-poisson neural networks for multi-part Hamiltonian systems with symmetries. Preprint. arXiv:2408.16160; 2024.
67. Vaquero M, de Diego DM, Cortés J. Designing poisson integrators through machine learning. Preprint. arXiv:2403.20139; 2024.
68. Vaquero M, Cortés J, de Diego DM. Symmetry preservation in Hamiltonian systems: simulation and learning. J Nonlin Sci 2024;34(6):115.
69. Tran HV. Hamilton–Jacobi equations: theory and applications, vol. 213. Providence: American Mathematical Society; 2021.
70. Lurie AI. Analytical mechanics. Berlin: Springer Science & Business Media; 2013.
71. Sanz-Serna JM. Symplectic integrators for Hamiltonian problems: an overview. Acta Numer 1992;1:243–86.
72. Leok M, Zhang J. Discrete hamiltonian variational integrators. IMA J Numer Anal 2011;31(4):1497–532.

73. Ohsawa T, Bloch, AM Leok M. Discrete Hamilton-Jacobi theory and discrete optimal control. In: 49th IEEE conference on decision and control (CDC). Piscataway: IEEE; 2010. p. 5438–43.

74. Ohsawa T, Bloch AM, Leok M. Discrete Hamilton–Jacobi theory. SIAM J Control Optim 2011;49(4):1829–56.

75. Bressan A. Viscosity solutions of Hamilton-Jacobi equations and optimal control problems. Lecture notes; 2011.

76. de León M, Sardón C. Geometry of the discrete Hamilton–Jacobi equation: applications in optimal control. Rep Math Phys 2018;81(1):39–63.

77. Evans LC, Souganidis PE. Differential games and representation formulas for solutions of Hamilton-Jacobi-Isaacs equations. Indiana Univ Math J 1984;33(5):773–97.

78. Elliott RJ. Viscosity solutions and optimal control. Pitman research notes in mathematics series. New York: John Wiley and Sons, Inc; 1987.

79. Bardi M, Dolcetta IC, et al. Optimal control and viscosity solutions of Hamilton-Jacobi-Bellman equations, vol. 12. Berlin: Springer; 1997.

80. Osher S, Fedkiw RP. Level set methods: an overview and some recent results. J Comput Phys 2001;169(2):463–502.

81. Rochet J-C. The taxation principle and multi-time Hamilton-Jacobi equations. J Math Econ 1985;14(2):113–28.

82. Forsyth PA, Labahn G. Numerical methods for controlled Hamilton-Jacobi-Bellman PDEs in finance. J Comput Finance 2007;11(2):1.

83. Mukundan R. Quaternions. Advanced methods in computer graphics: with examples in OpenGL; 2012. p. 77–112.

84. Vince J. Quaternions for computer graphics. Berlin: Springer; 2011.

85. Chirikjian GS. Information theory on lie groups and mobile robotics applications. In: 2010 IEEE international conference on robotics and automation. Piscataway: IEEE; 2010. p. 2751–7.

86. Chirikjian GS. Stochastic models, information theory, and lie groups, volume 1: classical results and geometric methods. Berlin: Springer Science & Business Media; 2009.

87. Chirikjian GS. Stochastic models, information theory, and Lie groups, volume 2: analytic methods and modern applications, vol. 2. Berlin: Springer Science & Business Media; 2011.

88. Hemingway EG, O'Reilly OM. Perspectives on Euler angle singularities, gimbal lock, and the orthogonality of applied forces and applied moments. Multibody Sys Dyn 2018;44:31–56.

89. Kozlov VV. The nonexistence of an additional analytic integral in the problem of the motion of a nonsymmetric heavy solid body around a fixed point. Vestnik Moskov Univ Ser I: Mat Mekh 1975;30:105–110.

90. Ziglin SL. Branching of solutions and the nonexistence of first integrals in Hamiltonian mechanics. II. Funct Anal Its Appl 1983;17(1):6–17.

91. Shrivastava SK, Modi VJ. Satellite attitude dynamics and control in the presence of environmental torques-a brief survey. J Guid Control Dyn 1983;6(6):461–71.

92. Markley FL, Crassidis JL, Markley FL, Crassidis JL. Attitude control. Fundamentals of spacecraft attitude determination and control 2014. p. 287–343.

93. Xie Y, Lei Y, Guo J, Meng B. Spacecraft dynamics and control. Berlin: Springer; 2022.

94. Marsden JE. Geometric foundations of motion and control. In: Motion, control, and geometry: proceedings of a symposium, board on mathematical science, National Research Council Education. Washington: National Academies Press; 1997.

95. Marsden JE, Ostrowski J. Symmetries in motion: geometric foundations of motion control. Nonlinear Sci. Today 1998:1–21.

96. Holm DD, Marsden JE, Ratiu TS. The Euler–Poincaré equations and semidirect products with applications to continuum theories. Adv Math 1998;137(1):1–81.

97. Kozlov VV. Integrability and non-integrability in hamiltonian mechanics. Russ Math Surv 1983;38(1):1.

98. Holmes P, Jenkins J, Leonard NE. Dynamics of the kirchhoff equations I: coincident centers of gravity and buoyancy. Phys D: Nonlin Phenom 1998;118(3–4):311–42.

99. Graver JG, Leonard NE. Underwater glider dynamics and control. In: 12th international symposium on unmanned untethered submersible technology. Citeseer; 2001.

100. Shashikanth BN. Dynamically coupled rigid body-fluid flow systems. Berlin: Springer; 2021.

101. Karapetian AV. On realizing nonholonomic constraints by viscous friction forces and celtic stones stability. J Appl Math Mech 1981;45(1):30–6.

102. Kozlov VV. The problem of realizing constraints in dynamics. J Appl Math Mech 1992;56(4):594–600.

103. Bloch AM. Nonholonomic mechanics. Berlin: Springer; 2015.

104. Bloch AM, Marsden JE, Zenkov DV. Nonholonomic dynamics. Not AMS 2005;52(3):320–9.

105. Bloch AM, Krishnaprasad PS, Marsden JE, Murray RM. Nonholonomic mechanical systems with symmetry. Arch Ration Mech Anal 1996;136:21–99.

106. Bloch AM, Marsden JE, Zenkov DV. Quasivelocities and symmetries in non-holonomic systems. Dyn Syst 2009;24(2):187–222.

107. Kozlov VV. On the integration theory of equations of nonholonomic mechanics. Preprint. nlin/0503027; 2005.

108. Gzenda V, Putkaradze V. Integrability and chaos in figure skating. J Nonlin Sci 2020;30(3):831–50.

109. Garcia JS, Ohsawa T. Controlled lagrangians and stabilization of Euler–Poincaré equations with symmetry breaking nonholonomic constraints. J Nonlin Sci 2024;34(5):91.

110. Hamel G. Über nichtholonome systeme. Math Ann 1924;92(1):33–41.

111. Hamel G. Theoretische Mechanik: eine einheitliche Einführung in die gesamte Mechanik, vol. 57. Berlin: Springer; 2013.

112. Ball KR, Zenkov DV, Bloch AM. Variational structures for Hamel's equations and stabilization. IFAC Proc Vol 2012;45(19):178–83.

113. Zenkov DV. On Hamel's equations. Theor Appl Mech 2016;43(2):191–220.

114. Shi D, Zenkov DV, Bloch AM. Hamel's formalism for classical field theories. J Nonlin Sci 2020;30:1307–53.

115. Fedorov YN, Zenkov DV. Discrete nonholonomic LL systems on lie groups. Nonlinearity 2005;18(5):2211.

116. Kobilarov M, De Diego DM, Ferraro S. Simulating nonholonomic dynamics. SeMA J 2010;50:61–81.

117. An Z, Gao S, Shi D, Zenkov DV. A variational integrator for the Chaplygin–Timoshenko Sleigh. J Nonlin Sci 2020;30:1381–419.

118. Soltakhanov SK, Yushkov M, Zegzhda S. Mechanics of non-holonomic systems: a new class of control systems. Berlin: Springer Science & Business Media; 2009.

119. Osborne JM, Zenkov DV. Steering the chaplygin sleigh by a moving mass. In: Proceedings of the 44th IEEE conference on decision and control. Piscataway: IEEE; 2005. p. 1114–1118.

120. Rhodes M, Putkaradze V. Trajectory tracing in figure skating. Nonlin Dyn 2022;110(4):3031–44.

121. Poincaré H, et al. Sur une forme nouvelle des équations de la mécanique. CR Acad Sci 1901;132:369–71.

122. Krishnaprasad PS, Marsden JE. Hamiltonian structures and stability for rigid bodies with flexible attachments. Arch Ration Mech Anal 1987;98:71–93.

123. Camassa R, Holm DD. An integrable shallow water equation with peaked solitons. Phys Rev Lett 1993;71(11):1661.

124. Gay-Balmaz F, Putkaradze V. Exact geometric theory for flexible, fluid-conducting tubes. CR Mec 2014;342(2):79–84.

125. Gay-Balmaz F, Putkaradze V. On flexible tubes conveying fluid: geometric nonlinear theory, stability and dynamics. J Nonlin Sci 2015;25:889–936.

126. Farkhutdinov T, Gay-Balmaz F, Putkaradze V. Geometric variational approach to the dynamics of porous medium, filled with incompressible fluid. Acta Mech 2020;231(9):3897–924.

127. Bondar DI, Gay-Balmaz F, Tronci C. Koopman wavefunctions and classical–quantum correlation dynamics. Proc R Soc A 2019;475(2229):20180879.

128. Gay-Balmaz F, Tronci C. Evolution of hybrid quantum–classical wavefunctions. Phys D: Nonlin Phenom 2022;440:133450.
129. Gay-Balmaz F, Tronci C. Reduction theory for symmetry breaking with applications to nematic systems. Phys D: Nonlin Phenom 2010;239(20–22):1929–47.
130. Gay-Balmaz F, Ratiu TS, Tronci C. Equivalent theories of liquid crystal dynamics. Arch Ration Mech Anal 2013;210(3):773–811.
131. Ellis DCP, Gay-Balmaz F, Holm DD, Putkaradze V, Ratiu TS. Symmetry reduced dynamics of charged molecular strands. Arch Ration Mech Anal 2010;197:811–902.

Index

A

Action, 2, 8, 29–32, 46, 48, 60, 80, 82, 104, 111–113, 115, 121, 132, 143, 144, 193

Action-angle variables, 110–115, 194

Adiabatic invariants, 112–115, 194

C

Canonical transformation, 75, 79, 86, 87, 89–103, 111, 113, 156, 190, 194

Chaplygin's sleigh, 136, 138

Configuration manifolds, 23–44, 46, 56, 76, 79, 81, 141, 142, 147–149, 152–155, 162, 164, 167, 171, 172, 176, 177, 181

Conservation of linear and angular momentum, 3–4, 6, 47

Conservation of phase volume, 99–101

Conserved quantities, 18, 45, 47, 72, 73, 137

Cotangent bundle, 70–71, 74, 79

Critical action principle, 30–32, 46

D

Darboux's and Liouville's theorem, 99–101

Differential forms, 75, 79–89, 94

Dissipative systems, 43, 49, 148, 172, 174

E

Energy conservation, 3, 4, 14, 19, 32, 33, 124, 133, 135

Equations, 2, 11, 25, 45, 55, 65, 79, 89, 103, 119, 131, 141, 147, 161

Equilibria, 34, 36, 43, 53–64, 112, 136, 147, 149, 150, 155, 173, 178

Euler, 7, 119, 121–125, 128, 129, 141, 144

Euler-Lagrange equations, 11–43, 45, 46, 48, 50, 55, 56, 65, 126, 141, 143, 147, 148, 152–155, 161, 163, 164, 166, 169–176, 198

Euler-Poisson equations, 125, 126, 130, 145

Euler's equation for a rigid body motion, 124

Euler top, 125, 128

Examples of nonholonomic systems, 138

Exterior calculus, 79–88, 186

F

Foundation of mechanics, 1, 42, 87

G

Generating functions, 94–99, 102, 103, 106, 107, 110, 113, 151, 156, 188, 194

H

Hamiltonian systems, 63, 65–77, 102, 104, 111, 114, 129, 155

Hamilton-Jacobi equations, 75, 79, 99, 103–115, 156, 189, 191

Holonomic constraints, 38–40, 131

I

Integrable cases of Euler-Poisson equations, 126, 129

© The Author(s), under exclusive license to Springer Nature Switzerland AG 2025
V. Putkaradze, *A Concise Introduction to Classical Mechanics*,
Surveys and Tutorials in the Applied Mathematical Sciences 16,
https://doi.org/10.1007/978-3-031-84977-0

Integral invariants, 79, 86–87, 95, 101, 113
Integration of differential forms, 86–87

K
Kepler's laws of planetary motion, 16, 18, 20
Kovalevskaya top, 126

L
Lagrange-d'Alembert's principle, 40–42,
    132–133, 135–137, 139, 157, 197
Lagrange top, 126–128
Legendre transform, 66, 67, 76, 150, 153, 155
Lie groups, 79, 119, 141, 142
Linear algebra, 53–55, 57, 59
Linearization, 164, 170
Liouville integrability conditions, 111, 112

M
Manifolds, 14, 23–43, 46, 56, 70, 72, 75, 76,
    79–82, 85, 86, 99, 112, 134, 141,
    142, 147–149, 152–155, 162, 164,
    167, 171, 172, 176, 177, 181, 182
Motion of central potential, 16–21, 49

N
Newton's laws of motion, 1–9
Noether's theorem, 4, 45–51, 139, 174
Nonholonomic constraints, 131–139

P
Poisson brackets, 71–75, 77, 85, 86, 89–93,
    155, 187

R
Rayleigh's dissipation function, 40–42, 44, 49,
    148, 154, 171, 173, 174
Rigid body, 117–130, 136, 139, 141–145, 196
Rigid body motion, 124
Rotation matrices, 47, 51, 117–119, 121, 129,
    141, 195

S
Small oscillations, 53–64, 112
Solutions of Hamiltonian systems, 73, 86, 87
Stability and instability, 59, 63
Symmetry of the Lagrangian, 45, 127
Symmetry reduction, 142
Systems with changing mass, 6, 8

T
Tangent bundle, 14, 26–30, 33–35, 37, 70, 74,
    80

V
Validity of nonholonomic constraints, 131–132
Variational principles in the presence of
    symmetry, 39, 40, 132, 133,
    143–145
Vector fields, 75, 79–84, 87, 88, 99, 150

W
Work and kinetic energy, 3